SpringerBriefs in Reproductive Biology

For further volumes:
http://www.springer.com/series/11053

Lee B. Smith · Rod T. Mitchell
Iain J. McEwan

Testosterone: From Basic Research to Clinical Applications

 Springer

Lee B. Smith
Rod T. Mitchell
MRC Centre for Reproductive Health
Queens Medical Research Institute
  University of Edinburgh
Edinburgh
UK

Iain J. McEwan
School of Medical Sciences,
  Institute of Medical Sciences
University of Aberdeen
AB
UK

ISSN 2194-4253   ISSN 2194-4261   (electronic)
ISBN 978-1-4614-8977-1   ISBN 978-1-4614-8978-8   (eBook)
DOI 10.1007/978-1-4614-8978-8
Springer New York Heidelberg Dordrecht London

Library of Congress Control Number: 2013948370

Printed on acid-free paper

Springer is part of Springer Science+Business Media (www.springer.com)

I don't mean to be crude, but it appears that testosterone basically has two, and only two, major drives: f**k it or kill it

Ken Wilber in 'A Brief History of Everything'

Preface

Is it a boy or a girl? That is the first question all new parents are asked on the arrival of a new baby. Whilst it could be argued that there is little other information that could be requested on the birth of a child, this neatly highlights the significance of gender, which will impact on the remainder of the child's life, including risk of specific disease and ultimately life expectancy.

The sex of the fetus is defined by the inheritance or not of a single gene, SRY, which is located on the Y chromosome. This gene is activated during fetal life to define Sertoli cells that drive the undifferentiated gonad towards a testicular fate. Steroid hormone producing Leydig cells develop within this fetal testis and begin producing testosterone which acts to masculinize the rest of the fetus, promoting the development of the male body form, including the penis, seminal vesicles, and prostate. Thus testosterone is the key factor that defines gender (in terms of our external body-form), and as such, directs our opportunities, restricts our lifestyle choices and influences our fate from our very beginnings.

In this volume we provide a brief résumé of the history of testosterone research, from the early pioneers through to the most recent advances in the field. We discover how steroid hormones were first identified and how testosterone was shown to be essential for male development. Moving forward we explore how and where testosterone is produced, and how the body controls testosterone production. We then investigate the impact testosterone has on different body systems both during their development and function, and how perturbation of testosterone action is associated with disease. We complete our story with an exploration of the emerging roles of testosterone in clinical therapy, and the future potential for manipulation of the testosterone signaling system for human health benefit.

Our aim in writing this volume is to arouse the curiosity of the reader and encourage them to delve further into specific areas of interest highlighted in this book. Whilst we have attempted to cover the majority of relevant aspects, the field is so diverse there will inevitably be certain aspects we are only able to touch lightly upon. Furthermore, as the remit of this book is to provide an introduction to the subject area for non-specialist readers, and to maintain clarity of the prose, we have shied away from inclusion of, or reference to, every possible primary research paper, focusing instead on key papers and reviews that, again, provide a point of access to each topic covered. We sincerely hope this book provides a stimulating read, containing as it does some novel and interesting facts regarding the wide-ranging roles of testosterone. We hope by the end to convince the reader that while testosterone is indeed a powerful chemical, its actions in the body are more varied and subtle than Ken Wilbur would have us believe.

June 2013   Lee B. Smith
                                                                                   Rod T. Mitchell
                                                                                   Iain J. McEwan

Contents

Abstract

Testosterone has long been considered the 'male hormone', though a deeper examination of the research into this molecule reveals its roles to be far more subtle and complex. In this volume we provide a brief résumé of the history of testosterone research, from the early pioneers through to the most recent advances in the field. We discover how steroid hormones were first identified and how testosterone was shown to be essential for male development. Moving forward we explore how and where testosterone is produced, and how the body controls testosterone production. We then investigate the impact testosterone has on different body systems both during their development and function, and how perturbation of testosterone action is associated with disease. We complete our story with an exploration of the emerging roles of testosterone in clinical therapy, and the future potential for manipulation of the testosterone signaling system for human health benefit. Together these chapters provide an entertaining and informative foray into the secretive world of testosterone.

Chapter 1
History of Masculinization

1.1 From the Ancient World to the Nineteenth Century

The establishment of our current understanding of masculinization has spanned centuries (Fig. 1.1). At the turn of the nineteenth century there was a growing obsession with the testes as a source for the "fountain of youth". In 1889, the German physiologist Brown-Séquard (1817–1894) added fuel to this fire, by claiming that injections of testicular extracts from guinea pigs and dogs were rejuvenating and restored vigor. Indeed this was a personal recommendation as Brown-Séquard administered the extracts to himself (Brown-Sequard 1889; Freeman et al. 2001); however, this was not something new.

Observations that the testes were important for masculinization and were thought to cure ailments in men were known from ancient times. In India, in the fifteenth century BCE, consuming the testes of animals was recommended as a treatment for impotence. While in ancient Greece, Aristotle wrote that castrating immature male birds affected secondary sexual characteristics, and in 'On the Generation of Animals', he noted the loss of the male phenotype after castration. In the mid-eighteenth century the Scottish surgeon John Hunter (1728–1793) experimented by transplanting the testes of a rooster into the bird's abdomen or the abdomen of a hen, where the transplanted tissue flourished with new blood vessels. At the time when medical practitioners could do more harm than good, John Hunter stood apart. He favored practical observations to understand how the body worked and was both an inspiring teacher and a skilled anatomist. He honed his surgical expertise by dissected human cadavers, a ready supply of which required murkier dealings with so-called Resurrectionists or more prosaically 'body-snatchers' (Moore 2005). In addition to his work on transplantations Hunter also made important observations on the tubular structures of the testes and correctly deduced that descent of the testes occurred in utero.

Given the ground breaking experimental work of John Hunter it is not too surprising that by the mid-nineteenth century Arnold Berthold (1801–1863) was able to reverse the behavioral and physiological effects of castration in roosters by transplanting the testes into the abdomen (Berthold 1944; Freeman et al. 2001;

L. B. Smith et al., *Testosterone: From Basic Research to Clinical Applications*,
SpringerBriefs in Reproductive Biology,
DOI: 10.1007/978-1-4614-8978-8_1, © The Author(s) 2013

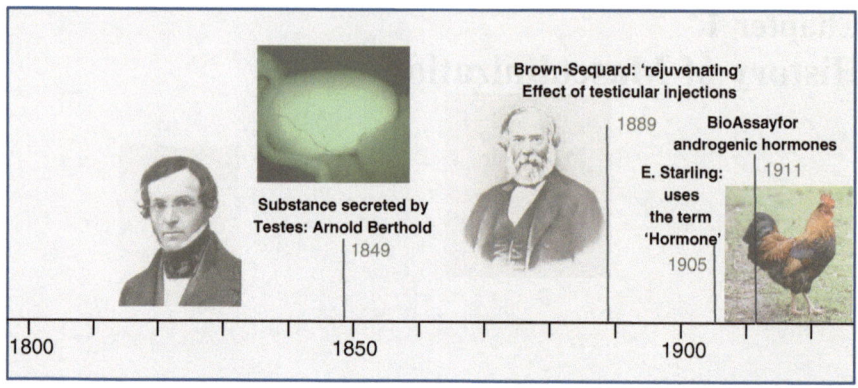

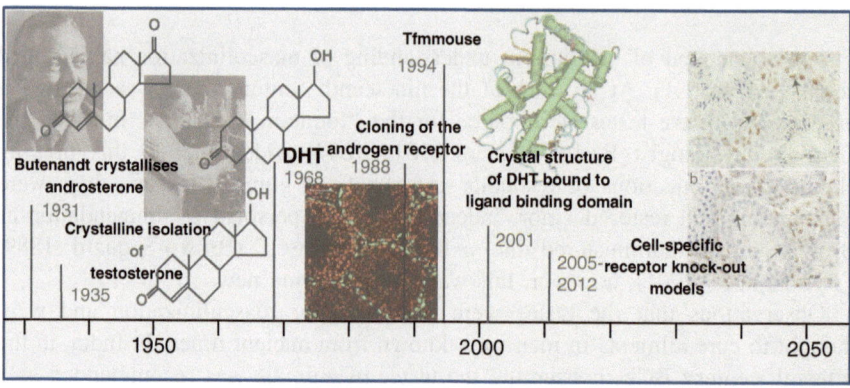

Fig. 1.1 The historical timeline of masculinization. A timeline detailing the major break-throughs underpinning our current understanding of masculinization, and the role played by androgens in this process

Miller and Fulmer 2007). However, these experiments were significant not just for confirming ancient wisdom, but because they revealed that the effect was most likely due to a substance being secreted from the transplanted tissues, as there were no nerves contacting the donor testes.

The concept of a substance secreted from a tissue into the blood circulation, and acting elsewhere in the body, was further developed by the French physiologist Claude Bernard (1813–1878) through his discovery that the liver is able to synthesize glucose, which is then secreted into the circulation. What we would now recognize as the endocrine system was defined by the studies of the British physiologists William Bayliss (1860–1924) and Ernest Starling (1866–1927), with their identification of "blood borne [chemical] messengers with targets far from the tissues of origin". This led to the first recorded use initially in a lecture by Starling, of the term 'hormone', to describe these chemical messengers; the word was derived from the Greek meaning 'I arouse to activity' and was suggested by Starling's colleague William Hardy (Freeman et al. 2001).

1.2 'Cell Therapy' or the Search for 'The Fountain of Youth'

Charles Brown-Séquard is perhaps best known for his studies with the testes, but he was also among the first to recognize the potential of secretions from thyroid, adrenals, pancreas, liver, and kidneys to treat diseases (Freeman et al. 2001). However, it was undoubtedly his personal testimony for the restorative benefits of testicular extracts that caught the imagination of the public as well as the medical establishment and led to the development of organotherapy. Although Brown-Séquard believed that aqueous extracts from young dogs and guinea pigs improved his physical and mental strength it is now accepted this was a placebo effect rather than a physiological property of the injected extracts, as the amount of testosterone present would have been several orders of magnitude lower than normal circulating levels (Cussons et al. 2002). However, this did not hinder the marketing of testicular extracts as an elixir of life capable of treating anything from epilepsy, paralysis, migraine, hysteria, arteriosclerosis, anemia to the flu (Fig. 1.2) (Hoberman and Yesalis 1995; Miller and Fulmer 2007).

THE STRAND MAGAZINE. 3

SEQUARINE

THE MEDICINE OF THE FUTURE.

THE one great remedy of the future will undoubtedly be the Serum. The mere fact that Scientists are now able to transfer energy from one animal body to another is sufficient to arouse enthusiasm among Doctors.

The perfection of the Sequarine Serum (which embodies the very essence of animal energy) in a form for everyday use, places animal therapy far in advance of other branches of medical science. This Serum is being used with astonishing success in treating :—

Nervousness,	Kidney Disease,	Paralysis,
Neurasthenia,	Diabetes,	Locomotor Ataxy,
Anæmia,	Dropsy,	General Weakness,
Rheumatism,	Dyspepsia,	Influenza,
Gout,	Liver Complaints,	Pulmonary
Sciatica,	Indigestion,	Troubles.

BROWN-SEQUARD,
F.R.S., F.R.C.P. (London),
who discovered the vital principle which is the basis of natural immunity from disease.

Fig. 1.2 Advertisement for sequarine 'the medicine of the future'. An advertisement, originally published in the British publication the *'Strand Magazine'* in 1912, extolling the supposed health benefits animal testicular extracts (*UCSF Digital Collections*, https://digital.library.ucsf.edu/items/show/1290)

The work of Brown-Séquard also led to the development of a number of surgical interventions as a means of rejuvenating aging men. The Viennese physiologist Eugen Steinach (1861–1944) pioneered the idea of autoplastic therapy, which involved vasoligation of the seminal ducts, which he believed would lead to an increase in hormone producing cells (Freeman et al. 2001; Miller and Fulmer 2007). This period, the early years of the twentieth century, also saw the rise in testicular transplants involving either chimpanzee or human tissue. In Europe, this rejuvenation operation was pioneered by the Russian surgeon Serge Voronoff (1866–1951) and in the USA by Frank Lydston (1858–1923) (Schultheiss and Engel 2003; Miller and Fulmer 2007; Kozminski and Bloom 2012). Lydston implanted the whole testis, from a human cadaver, into the patient's scrotum. His published work reported the benefits of sex gland implantation for a wide range of conditions, including senility, arteriosclerosis, chronic eczema, psoriasis, and cryptorchidism (Schultheiss and Engel 2003); and like Brown-Séquard before him, it seems he underwent the operation himself. In contrast, Voronoff used slices of primate testes, which were grafted onto the testis capsule of patients. He claimed that hormonal secretion was maintained for up to 2 years before declining (Schultheiss and Engel 2003; Miller and Fulmer 2007; Kozminski and Bloom 2012).

Although popular in the decade after the First World War the surgical route to rejuvenation eventually declined. This was no doubt due in part to the, at best, conflicting scientific evidence for success, and the growing taint of 'quackery' involving the use of animal testicular tissue and extracts, but ultimately because of the arrival in 1935 of a new elixir of life in the chemical shape of the androgen hormone testosterone.

1.3 Steroid Hunters

Thus, while the efficacy and supposed benefits resulting from injecting testicular extracts can be largely dismissed, the fact that the testes secreted a powerful chemical was inescapable. The early twentieth century was a golden age for steroid biochemistry, with many laboratories in both academia and the embryonic pharmaceutical industry actively engaged in purifying, characterizing, and synthesizing steroid hormones.

In 1911 McGee and co-workers established the capon's comb as a bioassay for 'androgenic substances' present in biological extracts, which involved measuring the growth of the comb and provided a robust quantitative assay (reviewed by Koch 1938; Freeman et al. 2001). In 1931, the German biochemist Adolf Butenandt (1903–1995) purified and identified the androgen androsterone (Fig. 1.3), from several thousand liters of human urine. Previously, this group had obtained the first pure steroid, estrone, from the urine of pregnant women. Working in Zurich Leopold Ruzicka (1887–1976) and co-workers achieved the chemical synthesis of androsterone (reviewed in Freeman et al. 2001). 1935 proved to be a landmark year, with Ernst Laqueur's (1880–1995) group at the University of Amsterdam purifying

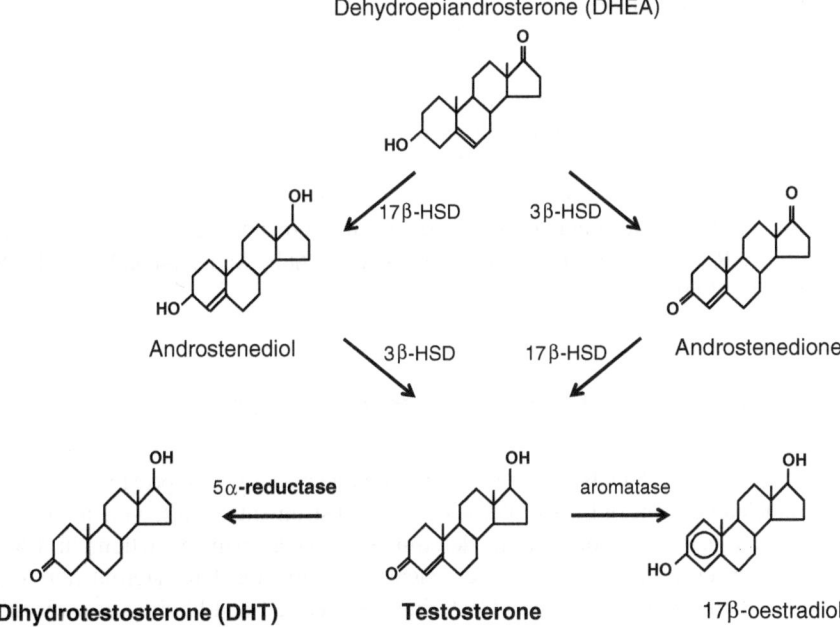

Fig. 1.3 Biosynthesis and conversion of testosterone. Testosterone, and other steroid hormones, are synthesized from cholesterol. This figure shows part of the biosynthetic pathway for testosterone from the androgen dehydroepiandrosterone (DHEA) and its metabolism to the more potent androgen, dihydrotestosterone (DHT) or the estrogen, 17β-estradiol. *HSD* hydroxysteroid dehydrogenase enzymes

testosterone from extracts of bull's testes and the groups of Butenandt and Ruzicka demonstrating the chemical synthesis of this hormone (Fig. 1.3) (reviewed by Koch 1938; Freeman et al. 2001); achievements that were recognized by the award of the Nobel Prize for medicine and physiology in 1939 to Butenandt and Ruzicka.

1.4   Dihydrotestosterone a Potent Androgen

Androgens have profound effects on the regulation of protein synthesis and the accumulation of body protein (Kochakian and Murlin 1935). It was investigations of this physiological action in the 1960 s that led to the discovery of the more potent metabolite of testosterone, dihydrotestosterone (DHT) (Fig. 1.3) by the Wilson laboratory at the University of Texas South-western Medical Center (Bruchovsky and Wilson 1999). During their studies to isolate bound radioactive steroid from the nucleus of rat ventral prostate, they were surprised to find DHT and not testosterone (Bruchovsky and Wilson 1968a). DHT is produced from testosterone by the enzyme 5α-reductase, which is present in peripheral androgen target tissues. Although DHT is thought to act primarily in an autocrine fashion, DHT is synthesized by the testes and low levels are found in the circulation so it can also function as a classical hormone.

Having identified these potent hormones, one of the outstanding questions was to understand how the hormone transmits its effect to target cells. A further consequence of the work from the Wilson laboratory, together with experiments from Liao and co-workers, was the development of the concept of a receptor protein present in the nucleus of the rat ventral prostate, termed the androgen receptor (AR; NR3C4) (Bruchovsky and Wilson 1968b; Anderson and Liao 1968; Fang et al. 1969).

1.5 Receptor Model for Steroid Hormone Action

Once testosterone had been purified and synthesized, it was possible to make radiolabeled versions of the steroid, which were instrumental in allowing molecular endocrinologists and biologists to determine its mechanism of action. In 1966, Elwood Jensen proposed the now classical two-step model for steroid hormone action, which involves the steroid entering a target cell and binding to a specific intracellular protein, termed a receptor. The receptor-hormone complex then moves to the nucleus and binds to DNA sequences in the genome to regulate gene expressions. Over the past 50 years the model has been refined and tailored to individual steroid receptors, but the underlying principal has stood the test of time. In the late 1980 s, the rodent and human cDNAs for the androgen receptor (AR) were cloned independently by several laboratories (Chang et al. 1988; Lubahn et al. 1988; Trapman et al. 1988). It is now estimated that the receptor has been cloned from one or more members of all the major vertebrate phyla; including a host a mammalian species, amphibians, birds, and fish. The cloning of the AR cDNA was a significant achievement that opened up the field of testosterone research to rigorous and in-depth investigation at the molecular, cellular, and tissue levels.

1.6 Key Points-Summary

- There has been a long and colorful history of the powerful effects of androgens on male physiology.
- A key breakthrough came with the isolation, molecular identification, and chemical synthesis of testosterone and other androgens in the early twentieth century.
- A further major development was the identification and cloning of the androgen receptor. This has paved the way for the comprehensive understanding how this hormone acts to regulate both male and female physiology.

References

Anderson KM, Liao S (1968) Selective retention of dihydrotestosterone by prostatic nuclei. Nature 219:277–279

Berthold AA (1944) The transplantaion of testes. B Hist Med 16:399–401

Brown-Sequard CE (1889) The effects produced on man by subcutaneous injections of a liquid obtained from the testicles of animals. Lancet 134:105–107

Bruchovsky N, Wilson JD (1968a) The conversion of testosterone to 5-alpha-androstan-17-beta-ol-3-one by rat prostate in vivo and in vitro. J Biol Chem 243(8):2012–2021

Bruchovsky N, Wilson JD (1968b) The intranuclear binding of testosterone and 5-alpha-androstan-17-beta-ol-3-one by rat prostate. J Biol Chem 243(22):5953–5960

Bruchovsky N, Wilson JD (1999) Discovery of the role of dihydrotestosterone in androgen action. Steroids 64:753–759

Cussons AJ, Bhagat CI, Fletcher SJ, Walsh JP (2002) Brown-Sequard revisited: a lesson from history on the placebo effect of androgen treatment. Med J Aust 177:678–679

Chang CS, Kokontis J, Liao ST (1988) Molecular cloning of human and rat complementary DNA encoding androgen receptors. Science 240:324–326

Fang S, Anderson KM, Liao S (1969) Receptor proteins for androgens. On the role of specific proteins in selective retention of 17-beta-hydroxy-5-alpha-androstan-3-one by rat ventral prostate in vivo and in vitro. J Biol Chem 244:6584–6595

Freeman ER, Bloom DA, McGuire EJ (2001) A brief history of testosterone. J Urol 165:371–373

Hoberman JM, Yesalis CE (1995) The history of synthetic testosterone. Sci Am 272(2):76–81

Koch FC (1938) The chemistry and biology of male sex hormones. Bull N Y Acad Med 14:655–680

Kochakian CD, Murlin JR (1935) The effect of male hormone on the protein and energy metabolism of castrate dogs. J Nutr 10(4):437–459

Kozminski MA, Bloom DA (2012) A brief history of rejuvenation operations. J Urol 187:1130–1134

Lubahn DB, Joseph DR, Sar M et al (1988) The human androgen receptor: complementary deoxyribonucleic acid cloning, sequence analysis and gene expression in prostate. Mol Endocr 2:1265–1275

Moore W (2005) The knife man: blood, body-snatching and the birth of modern surgery. Bantum Press, Great Britain, 639 pp.

Miller NL, Fulmer BR (2007) Injection, ligation and transplantation: the search for the glandular fountain of youth. J Urol 177:2000–2005

Schultheiss D, Engel RMG (2003) Frank Lydston (1858-1923) revisited: androgen therapy by testicular implantation in the early twentieth century. World J Urol 21:356–363

Trapman J, Klaassen P, Kuiper GG et al (1988) Cloning, structure and expression of a cDNA encoding the human androgen receptor. Biochem Biophys Res Com 153:241–248

Chapter 2
Recent Understanding of Masculinization

2.1 The History of Masculinization

From ancient times, man has sought to understand what underlies the divergent body forms of men and women. Aristotle (ca 350 BCE) theorized that the semen determines the form of the embryo, and is the driving force towards maleness, whereas the female provides the nurturing environment; sex determination is decided upon by a battle between these two factors for supremacy (Josso 2008). Whilst this is not how sex determination works, Aristotle was surprisingly close to the mark in terms of suggesting the existence of a male factor that determines the gender of the fetus.

However it would be another 2000 years before our understanding was further advanced. In the 19th century, scientists had recognized that sexual differentiation of the fetus occurred only after an indifferent stage, when the gonads, internal reproductive structures and external genitalia are identical in males and females. A breakthrough came in 1903 when Bouin and Ancel identified the testicular cells responsible for male characteristics, and suggested that a substance (the term 'hormone' had not been invented at this time) secreted by these cells drove masculinization (Bouin and Ancel 1903).

The hormonal theory of sex determination was established in a series of groundbreaking papers in 1916, using Freemartin cattle, which derive from multiple pregnancies, involving at least one male fetus. These are genetically female cows that are, to a certain extent masculinzed in utero. At birth freemartin ovaries are reduced in size and the female reproductive system is either absent or underdeveloped. Conversely, some aspects of a male reproductive system for example seminal vesicles and epididymides are well developed. Both Lillie and Keller and Tandler independently suggested that this was the result of the exchange of some testicular secretion to the female fetus from a male twin, via vascular connections between placentas (Lille 1916; Keller and Tandler 1916).

The father of modern sexual differentiation is widely accepted to be Alfred Jost, whose contribution to science is excellently reviewed in (Josso 2008). In 1938 Jost joined the laboratory of Professor Robert Courrier who was exploring the roles of hormones in gestation. It was the coincident first availability of synthesized hormones (see Chap. 1) that was to permit Jost to fundamentally change our understanding of masculinization.

In 1947 Jost showed that castrating fetal rabbits prevented masculinization, thus demonstrating that the testis determines the phenotypic sex of the individual (Jost 1947). Not only this, but Jost identified that whatever the date of castration, all female rabbit fetuses retained their female traits, whilst castration of male fetuses had widely varying effects dependent upon day of castration. At day 23 or 24 male development proceeded normally, whereas castration at 19–21 days resulted in acquirement of a female form. Thus Jost had demonstrated the primacy of testicular hormones, that fetuses develop as female unless male hormones prevent them from doing so, and that timing of action of the male hormone(s) was critical to male development.

In further studies, Jost transplanted a fetal rabbit testis into a female rabbit fetus. This replicated the observations in male fetuses, that both male development and regression of the female developmental tract (the Müllerian duct) required testicular factors; it also established that female fetuses were competent to respond to these male factors. However, whilst substitution of the transplanted testis with a crystal of testosterone prioprionate promoted masculinization, it failed to cause Müllerian duct regression. Jost proposed a second factor; an Anti Müllerian Hormone (AMH or MIS) existed, and was secreted by the testis to promote regression of female reproductive primordia (Jost 1953). Despite much objection to this idea at the time, Jost was eventually proved correct, and Amh was identified as a member of the TGF-Beta family of secreted glycoproteins (Josso 1990). Knockout of Amh in mice resulted in retention of the female reproductive system in male mice, whilst transgenic expression of Amh in female fetuses drove, normally male-specific, Müllerian duct regression.

Thus the production of testosterone and Amh by the testis are the two driving factors controlling masculinization of the mammalian fetus (Fig. 2.1); but the production of testosterone and Amh themselves rely upon development of the architecture of the fetal testis.

2.2 The Ontogeny of Testicular Development

The fetal gonads arise as outgrowths from the mesenephros at approximately embryonic day 10 in mice, or 5–6 weeks gestation in humans. Each gonadal-ridge separates into two unequal sections. The most cranial, smaller section goes on to develop into the adrenal gland, whilst the larger caudal section goes forward into gonadal development. The germ cells are defined very early during development and migrate through the developing gut to take up position in the fetal gonad by embryonic day 10 in the mouse. Thus at embryonic day 10.5 in the mouse the gonad is phenotypically indistinguishable between male and female fetuses.

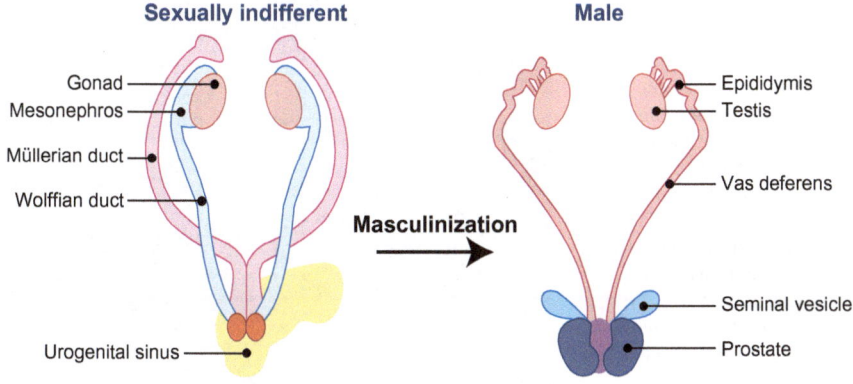

Fig. 2.1 *Masculinization of the reproductive system* Prior to testosterone and Amh action, the reproductive system is sexually indifferent with both Müllerian and Wolffian ducts developing within the mesonephros. Production of Amh by the testicular Sertoli cells promotes regression of the Müllerian duct (the precursor of the female reproductive system), while testosterone production by the fetal testis stabilizes and promotes development of the Wolffian duct into the prostate, vas deferens, seminal vesicles and epididymides (Modified from Welsh et al. (2008))

Inheritence of a Y-chromosome, and specifically inheritance of the HMG-box transcription factor Sry (which is located on the Y chromosome), drives development towards the testis lineage. Sry is expressed in cells around the periphery of the gonad, and activates Sox9, resulting in development of Sertoli cells around the periphery of the developing testis, which then migrate into the testis. The Sertoli cells begin to surround the germ cells, forming the testis cords, which then recruit cells of the testicular interstitium to take on a myoid fate, surrounding the testis cords in a thin sheath of smooth muscle. Sertoli cells also secrete Platelet Derived growth Factor-alpha (PDGF-α) and Desert hedgehog, which promote development of the fetal Leydig cells within the testicular interstitium, and by embryonic day 13.5 the testicular architecture and cell-types have all been established (Fig. 2.2). From embryonic day 13, Sertoli cells secrete Amh, which signals to the cognate Amh receptor located in the developing Müllerian duct—the anlagen of the female reproductive system, instructing it to regress. At the same time, the developing fetal Leydig cells initiate steroid hormone production (reviewed in Quinn and Koopman (2012)).

2.3 Human Conditions Associated with Androgen Disruption During Development

The spectrum of conditions associated with perturbation of androgen signaling during development is wide-ranging in both its severity and underlying causes, but can be crudely separated into two separate groupings, enclosed within the wider

Fig. 2.2 *The architecture of the fetal testis* Immunohistochemical staining of an embryonic day 21 fetal rat testis revealing the architecture of the testis. *Blue* = peritubular myoid cells, *red* = fetal Leydig cells, *Green* = interstitial cells. Note Sertoli and germ cells are found inside the peritubular myoid encompassed tubules (*black* in this image). Image kindly provided by Professor Richard Sharpe, University of Edinburgh

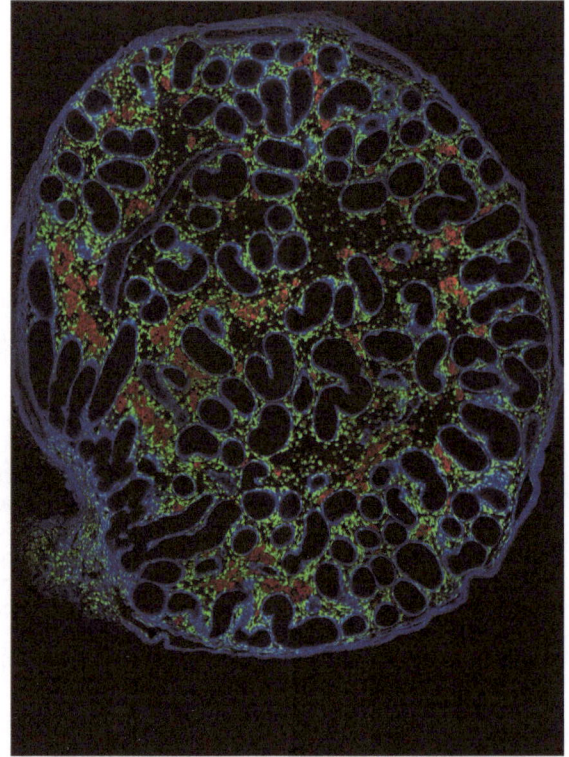

umbrella of disorders of sexual development (DSD). These are (1) boys that fail to complete the masculinization process, and (2) girls that undergo some level of masculinization.

Testosterone mediates the majority of its developmental functions by binding to its cognate receptor AR. Boys lacking a functioning AR, are unable to transmit the effect of testosterone to the cells of the body. As such, despite being genetically XY, these individuals develop a female body form and female genitals, yet retain internal testes. These individuals may go for years without realizing they are genetically male. However, despite not being able to respond to testosterone's masculinization signal, their testes still express and respond to Amh, and thus these individuals lack an internal female reproductive system. Diagnosis is often made during puberty, following investigation to establish the cause of failure of initiation of menarche (periods). This condition and is referred to as Complete Androgen Insensitivity Syndrome (CAIS; see Chap. 6) (Hughes 2012).

CAIS represents the form most divergent from the classic male phenotype, yet CAIS is simply at one end of a spectrum of androgen insufficiency disorders termed partial androgen insensitivity syndrome (PAIS); one of these results from a reduction in the ability to propagate the androgen signal. 5-alpha-reductase is produced in peripheral tissues to enhance the potency of the androgen signal by producing DHT (see Chap. 1), which has a higher affinity/lower dissociation rate for the AR. As such, as testosterone is diluted in the bloodstream conversion to DHT re-establishes its potency by essentially amplifying the signal approximately tenfold.

Individuals diagnosed with 5-alpha-reductase deficiency have a loss of function mutation in one of the genes coding for 5-alpha-reductase, which prevents conversion of testosterone to the more potent AR agonist DHT. As such they exhibit under-masculinisation phenotypes, with genitals phenotypically mid-way between male and female (Chong 2011).

At the milder end of the spectrum, two of the most common congenital abnormalities observed at birth in boys are hypospadias (Kalfa et al. 2011) and cryptorchidism (Hutson et al. 2013). Hypospadias is a condition where the closure of the urethral fold is perturbed such that the opening of the urethra is at the base, or along the shaft of the penis; whilst cryptorchidism is the failure of either one or both of the testes to descend into the scrotum. Both conditions are usually corrected surgically soon after birth, and rarely present any long-term impacts, although a correlation between cryptorchidism at birth and testicular cancer in adulthood has been observed in a minority of cases.

The most common condition driving aberrant masculinization in female fetuses is congenital adrenal hyperplasia (CAH) (Huynh et al. 2009). The adrenal functions in fetal life to produce cortisol. A mutation in any one of several of the enzymes within the steroid hormone synthesis pathway (usually Cyp21 or in rarer cases Cyp11b1) can suppress cortisol production. A reduction in cortisol leads to increased release of adrenocorticotrophic hormone (ACTH), which drives proliferation of the steroid producing cells of the adrenal. Because the enzymes required to make cortisol are not functioning, the steroid precursor molecules are diverted towards other steroid synthesis pathways, resulting in high levels of androgen production by the adrenal. In boys the increased androgens may result in virilisation in childhood, which may predispose to precocious puberty and the associated glucocorticoid and minealocorticoid deficiency can cause death in the neonatal period from a salt-wasting crisis. However in females, the unusually high concentrations of androgens result in masculinization of the reproductive system. Girls with CAH are often born with ambiguous genitalia, part way between male and female due to the virilising effects of androgens. This again confirms the observations of Jost, that the female fetus is competent to respond to androgens, and it is the absence of the androgen ligand that prevents masculinization in normal female development.

Jost's original observation, that masculinization could be affected only during specific stages of male development has been further established in recent times. One of the concepts to arise from this is the concept of 'fetal programming', where incidences occurring in fetal life have lifelong impacts. In the reproductive field this has led directly to the development of the testicular dysgenesis syndrome (TDS) hypothesis. This states that perturbation of androgen-mediated events in fetal life underpins many of the clinical conditions observed in later life. Whilst this is perhaps easy to accept for conditions present at birth such as hypospadias, emerging evidence suggests, perhaps incredibly, that this still holds true for conditions only evident in adulthood, for example, low sperm count or testicular germ cell cancer (Sharpe and Skakkebaek 2008).

2.4 The Universal Male Programming Window

The mechanistic relationship between subnormal androgen action in fetal life and these conditions has been firmly established in animal studies. Building on from Jost's original observations, in a series of groundbreaking studies Welsh and colleagues demonstrated the existence of a universal male programming window (MPW) a specific window in fetal life in which androgen action was required to specify development of male specific organs such as the seminal vesicles and prostate (Welsh et al. 2008) (Fig. 2.3). Importantly, they demonstrated that androgen action at any other time during fetal development or postnatal life was

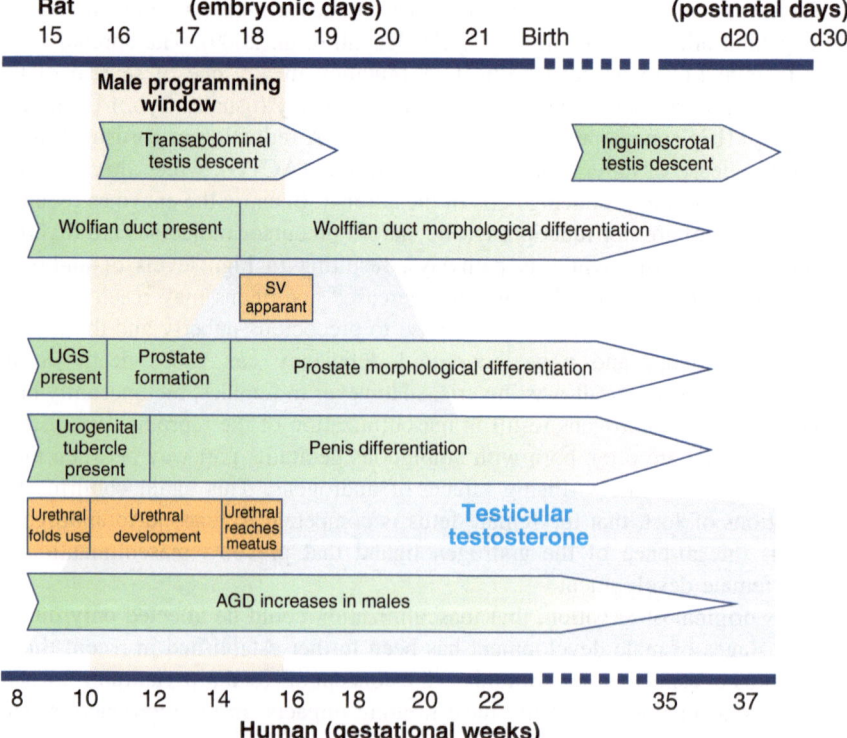

Fig. 2.3 *Identification of the Universal Male Programming Window* A schematic diagram showing the masculinization programming window and impact on reproductive tissues in rats and humans. S.V. = Seminal vesicles. (Modified from Welsh et al. (2008))

unable to rescue the failure of action in this programming window. The novelty of the studies was the realization that androgen action was required several days prior to morphological differentiation between the sexes, thus androgens programme male development in advance.

The simplest way to understand this concept is the case of testicular descent into the scrotum, failure of which, as we have already discussed, results in cryptorchidism. Welsh and colleagues demonstrated that androgen action between embryonic day 15 and 17 in the rat fetus was essential for testicular descent, which in the rat occurs in the third week after birth. Androgen action throughout the period between embryonic day 17 and 3 weeks post-partum was unable to rescue failure of androgen action during these two key days in fetal life.

Welsh and colleagues went on to show the programming window applied equally to Wolffian duct maturation, penis development, seminal vesicle development and prostate development (Fig. 2.3). Whilst androgens are required throughout development to ensure proper development, initial presence of the organ was wholly dependent upon androgen action at this key developmental stage (Welsh et al. 2008).

One of the surprising aspects of this study was the identification that ano-genital distance (AGD), which is known to be about twice as large in males as females was also programmed in the MPW. This led to the suggestion that AGD could provide a biomarker of androgen action during the specific time-point of the MPW. Whilst this has been shown to be true in rat models with or without blockade of androgen action, recent evidence suggests a similar relationship exists in humans (reviewed in Dean and Sharpe (2013)). One of the confounding factors though is that androgen action in the MPW only defines the maximal size any organ is able to grow. Androgen action after birth is required to reach this maximal size, but exogenous androgens cannot increase the size of any organ beyond that programmed in during development. This also suggests that interference in the androgen signal in postnatal life, for example exposure to estrogenic compounds, endocrine disruptors or obesity could prevent maximal development. Although research into this area is still ongoing, and exactly how plastic the developmental system is remains to be established. Manipulation with androgenic or estrogenic compounds is a current hot topic in this field of research.

2.5 Key Points-Summary

- Masculinisation requires Both Anti-Müllerian Hormone (Amh) and Testosterone, secreted by the fetal testis
- Both male and female fetuses are able to respond to Amh and Testosterone. It is the presence these ligands that drives masculinization
- Both Amh and Testosterone must act in a defined window to promote male development

- Testosterone action in this 'Male Programming Window' is essential for correct male development
- Perturbation of the MPW can lead to clinical pathologies with life-long impacts.

References

Bouin P, Ancel P (1903) Sur la signification de la gland interstitialle du testicule embryonnaire. Comptes-Rendus de la societe de biologie 55:1632–1634

Cheon CK (2011) Practical approach to steroid 5alpha-reductase type 2 deficiency. Eur J Pediatr 170:1Đ8

Dean A, Sharpe RM (2013) Anogenital distance or digit length ratio as measures of fetal androgen exposure: relationship to male reproductive development and its disorders. J Clin Endocrinol Metab 98(6):2230–2238. doi: 10.1210/jc.2012-4057. Epub 8 April 2013

Hughes IA, Davies JD, Bunch TI, Pasterski V, Mastroyannopoulou K, MacDougall J (2012) Androgen insensitivity syndrome. Lancet 380(9851):1419–1428

Hutson JM, Southwell BR, Li R, Lie G, Ismail K, Harisis G, Chen N (2013) The regulation of testicular descent and the effects of cryptorchidism. Endocr Rev (May 10. Epub ahead of print)

Huynh T, McGown I, Cowley D, Nyunt O, Leong GM, Harris M (2009) Cotterill AM The clinical and biochemical spectrum of congenital adrenal hyperplasia secondary to 21-hydroxylase deficiency. Clin Biochem Rev 30(2):75–86

Josso N (1990) Anti-mŸllerian hormone: hormone or growth factor? Prog Growth Factor Res 2(3):169–79 (Review)

Josso N (2008) Professor Alfred Jost: the builder of modern sex differentiation. Sex Dev 2:55–63

Jost A (1947) Recherches sur la differenciation sexuelle de lÕembryon de lapin. III. Role des gonads foetales dans la differenciation sexualle somatique. Arch Anat Mcrosc Morph Exp 36:271–315

Jost A (1953) Problems of fetal endocrinology: the gonadal and hypophyseal hormones. Rec Progr Horm Res 8:379–418

Kalfa N, Philibert P, Baskin LS, Sultan C (2011) Hypospadias: interactions between environment and genetics. Mol Cell Endocrinol 335:89–95

Keller K, Tandler J (1916) Uber das Verhalten der Eihaute bei derZwillingstrachtigkeit des Rindes. Untersuchungenuber die Entstehungsursache der geschlechlichen unterentwicklung von weiblichen zwillingskalbern, welche neben einem mannlichen kalbe zur Entwicklung gelangen. Wiener Tierartzl Monatschr 3:316–526

Lillie FR (1916) Theory of Freemartin. Science 43:611–613

Quinn A, Koopman P (2012) The molecular genetics of sex determination and sex reversal in mammals. Semin Reprod Med 30:351–363

Sharpe RM, Skakkebaek NE (2008) Testicular dysgenesis syndrome: mechanistic insights and potential new downstream effects. Fertil Steril 89(2 Suppl):e33–e38. doi:10.1016/j.fertnstert.2007.12.026

Welsh M, Saunders PT, Fisken M, Scott HM, Hutchison GR, Smith LB, Sharpe RM (2008) Identification in rats of a programming window for reproductive tract masculinization, disruption of which leads to hypospadias and cryptorchidism. J Clin Invest 118:1479–1490

Chapter 3
Where Does Testosterone Come from and How Does It Act?

3.1 Testes-Cellular Morphology

The adult testis contains several different cell types, each with a specific role to play in promoting the two functions of the testis,—(1) the production of sperm and (2) steroid hormones including testosterone (Fig. 3.1).

The testis is structurally split into two distinct areas: the seminiferous tubules and the interstitium. The seminiferous tubules are essentially sausage shaped-tubes, consisting of the developing germ cells and the Sertoli cells (which together form the seminiferous epithelium), which nurture and support the germ cells during sperm development. The mesoepithelial Sertoli cells have a polarity, and are attached to a basement membrane at the periphery of the tubules, with their apical processes located in the tubule lumen. A simple analogy is to think of the Sertoli cells as a tree, and the maturing sperm as apples, growing to ripeness in the branches (for review, see Skinner and Griswold 2005). Just like a tree the Sertoli cells, provide physical support, nutrient support, and protection from pathogens, to ensure production of viable sperm. Germ cells develop from spermatogonial stem cells found along with the basement membrane, undergoing a three-step process of development toward their eventual release into the lumen of the tubules. First they undergo a series of mitotic divisions to increase the pool of spermatogonia, these then undergo meiosis, to produce haploid cells containing just a single copy of each chromosome. Finally, these haploid germ cells undergo a process called spermiogenesis, where they lose the majority of their cytoplasm, package their DNA incredibly tightly and develop the characteristic shape of the sperm. Once this process is complete, the mature sperm cells are released from the seminiferous epithelium into the tubule lumen, and begin their journey toward the epididymis (process reviewed in O'Donnell and de Kretser 2013).

The testicular interstitium (the spaces between the seminiferous tubules) houses a number of cell types vital to testis function. Surrounding the seminiferous tubules are the peritubular myoid (PTM) cells. These are myofibroblast cells (one layer thick in rodents, several in humans), which form the "sheath of the sausage," and contract to aid sperm passage along the seminiferous tubule lumens. The testicular

L. B. Smith et al., *Testosterone: From Basic Research to Clinical Applications*,
SpringerBriefs in Reproductive Biology,
DOI: 10.1007/978-1-4614-8978-8_3, © The Author(s) 2013

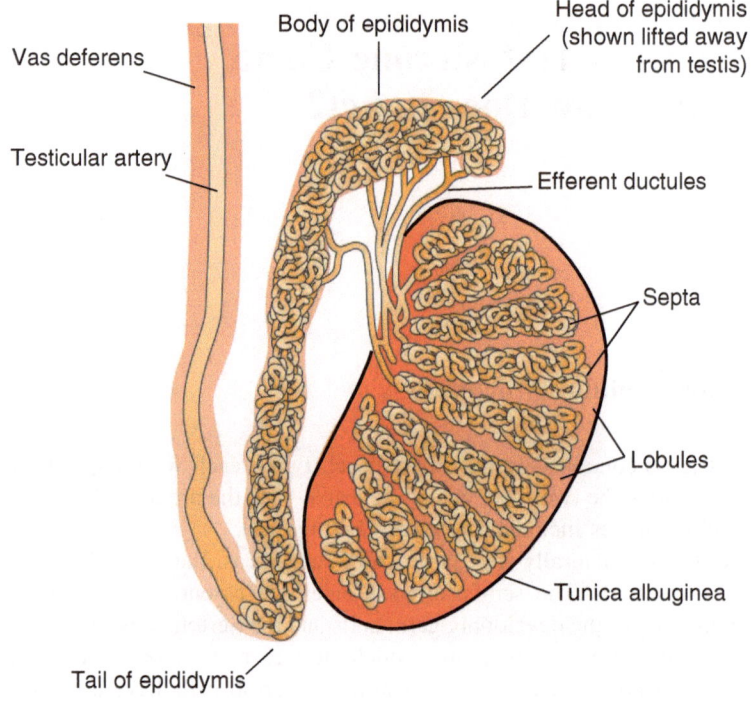

Fig. 3.1 Cell types of human testes. The testis is separated into two discrete regions; the seminiferous tubules, and interstitial spaces between the tubules. Seminiferous tubules are constructed of developing germ cells, associated with supporting Sertoli cells, which act to nurse germ cells during spermatogenesis. Surrounding the seminiferous tubules is a layer of smooth muscle PTM cells. The testosterone-producing Leydig cells are found in the interstitial spaces, along with the testicular vasculature, immune cells, and fibroblasts. The entire testis is surrounded by the tunica albuginea, a thick layer of smooth muscle cells

interstitium also contains a number of arterioles, capillaries, and venules that traverse the interstitial space, bringing nutrients and oxygen to the testis and carrying away testosterone to the rest of the body. In addition, the interstitium contains several lymph vessels, which provide a secondary mechanism for returning hormones to the rest of the body. Residing in between the seminiferous tubules and the vascular system are cells of the immune system, including resident macrophages, which closely interact with the steroid hormone producing Leydig cells (process reviewed in O'Donnell and de Kretser 2013).

Two populations of Leydig cells have been identified (for review of Leydig cells see Payne and Hardy 2007). The first population, the fetal Leydig cells develop in the fetal testis and persist until approximately postnatal day 20 in the mouse, when they are replaced by a population of Adult Leydig cells. These adult Leydig cells are believed to develop from precursors resident in the interstitium, which undergo a series of mitotic divisions, reaching their final number and maturity at approximately day 50. Whilst phenotypically distinguishable under electron microscopy, the similar roles of the two populations, production of the two hormones Insl3 and androgens, means that it is challenging to distinguish between the populations based upon immunohistochemical localization of specific proteins. Nevertheless, the take home message is that Leydig cells function as the androgen factory in males, making over 95 % of androgens, and as such are the key cell type in testosterone production.

3.2 Steroidogenesis

The substrate for production of all steroid hormones is Cholesterol. Production of steroid hormones from cholesterol is similar in all steroidogenic tissues, being largely dependent upon the presence and activity of specific steroidogenic enzymes. A good analogy for the process of Steroid hormone production is to imagine it as a series of stepped waterfalls, with product able to flow to the next step if the requisite enzyme is present. If the enzyme is absent then the substrate moves down an alternative route, via a different enzymatic pathway, resulting in an alternative product. For example, presence of the enzyme 21-hydroxylase in the adrenal gland permits production of cortisol from a progesterone precursor, whilst in the testis, 21-hydoxylase is absent, and the progesterone is instead converted to androgens by 17, 20, lyase. This is an important consideration when examining the Leydig cell products, for example—transgenic over-expression of aromatase, which converts testosterone to estrodiol, in the testis, not only increases available estrogen, but also suppresses testosterone concentrations by readily converting available testosterone to estrogens. Thus, changes in enzyme activity can have significant impacts in terms of which steroid hormones are produced, the results of which are associated with a variety of severe clinical pathologies (for extensive review, see Miller and Auchus 2011) (see Chap. 6).

Under normal conditions, testosterone production by Leydig cells is stimulated by binding of LH to LH receptors present on the cell surface. This initiates cAMP-mediated induction of Protein Kinase A production which is required for transport of cholesterol from the cytoplasm to the outer mitochondrial membrane. Steroidogenic acute regulatory protein (StAR) then transports the cholesterol across the mitochondrial membrane where the enzyme P450 side-chain cleavage (P450scc), which is resident on the inner mitochondrial membrane, converts cholesterol into pregnenolone. This is ultimately transferred to the smooth endoplasmic reticulum, where testosterone is synthesized by a series of steroidogenic enzymes such as Cyp11a1, Cyp17a1, HSD3B2, and HSD17B3. Testosterone Biosynthesis then follows either the Δ^4 pathway, via progesterone 17-hydroxy (17-OH) progesterone and androstenedione or the Δ^5 pathway via pregnenolone (17-OH) pregnenolone, and dehydroepiandosterone (DHEA). The Δ^4 pathway is primarily used in the mouse whilst the Δ^5 pathway is preferred in humans (for extensive review see Miller and Auchus 2011) (Fig. 1.3). The testosterone product is secreted into the testis and bloodstream, where it signals back at the level of the hypothalamus and pituitary to suppress LH secretion. Thus, a negative feedback loop is established, where low testosterone promotes LH-mediated stimulation of testosterone production by Leydig cells, and high testosterone suppresses the LH signal, reducing steroidogenesis in Leydig cells.

3.3 Adrenal Gland

Whilst androgens are the main source of testosterone in males, the adrenal gland is another important source of androgens in humans (though not in rodents) (reviewed in Sidiropoulou et al. 2012). Synthesis is under the control of Adrenocorticotrophic Hormone (ACTH), secreted from the pituitary, which stimulates steroid hormone synthesis in the adrenal. The primary function of ACTH is to stimulate glucocorticoid and mineralocorticoid production by the adrenals, however, expression of P450C17 in the zona fasiculata and reticularis, and specifically its 17, 20 lyase function, permits conversion of 17OH-pregnenolone to DHEA, and 17OH-progesterone to androstenedione, respectively. These androgens are not thought to play a significant role in males, apart from a role in adrenarche (the activation of the adrenal axis that occurs around the onset of puberty), however, in females excessive adrenal androgens through increased phosphorylation of P450c17 have been associated with increased hirsuitism and Poly Cystic Ovary Syndrome (PCOS), which is linked to female infertility (see Chap. 6).

3.4 Ovary

Although testosterone is often called the male sex hormone it is important to appreciate that the female gonads (ovaries) also produce and respond to testosterone. Under the control of LH the theca cells in the ovary synthesize and secrete testosterone (Hillier 2001; Jamnongjit and Hammes 2006). Furthermore, numerous cell types in the ovary contain the AR, including the surface epithelia, granulosa cells of developing follicles and stromal cells (Chadha et al. 1994; Hillier et al. 1997; Saunders et al. 2000; Weil et al. 1998).

Testosterone produced in the ovary in part represents a pro-hormone, a substrate which is metabolized by the aromatase enzyme to estradiol in granulosa cells in response to follicle stimulating hormone (FSH) (Hillier 2001; Jamnongjit and Hammes 2006). However, the presence of the mRNA and protein for the AR indicate a direct role for androgens in female reproductive physiology. This is further emphasized by mouse models, where the ovarian AR gene has been removed (knocked out): these animals develop normally but are sub-fertile, compared with litter-mate controls, indicating a reproductive defect (Hu et al. 2004; Shiina et al. 2006; Walters et al. 2008; Yeh et al. 2002).

3.5 The Androgen Receptor (NR3C4)

The gene for the AR is almost 200 kb in size and is located on the X-chromosome, at position Xq11-12, and is therefore a single copy gene in males. The coding sequence comprises of eight exons (Fig. 3.2). The highest expression of the mRNA transcript is in reproductive tissues such a prostate and uterus, but high levels of the receptor mRNA have also been described in non-reproductive tissues, for example liver and fat [www.nursa.org/index.cfm].

The availability of the receptor cDNA ushered in a new era of molecular and structural investigation of AR action. The human receptor has up to 919 amino acids and a molecular weight of 110 kDa. The exact number of amino acids varies due to polymorphisms in two stretches of the amino acids glutamine (Q) and glycine (G) present in the N-terminal domain (NTD) (Fig. 3.2) (reviewed in McEwan 2004).

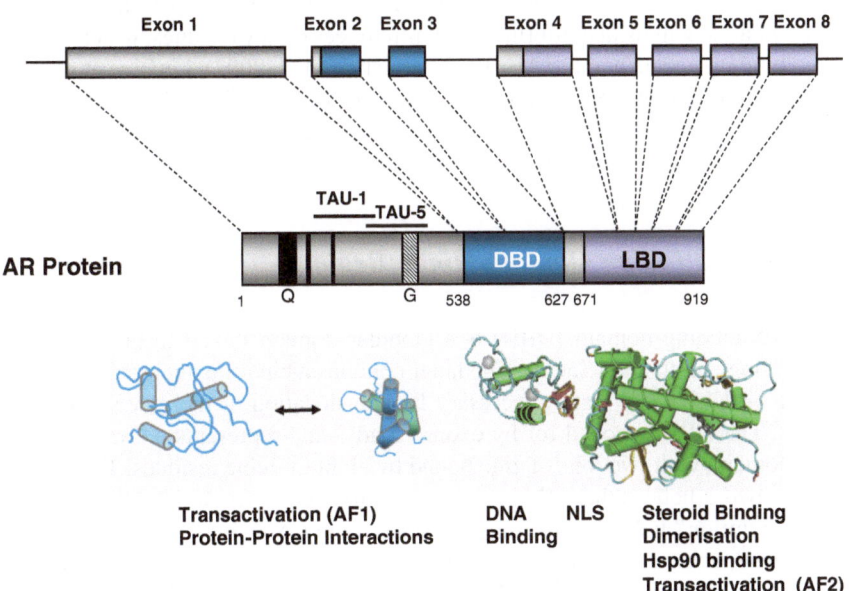

Fig. 3.2 Organization of the androgen receptor gene and protein. *Top*, shows the exon intron structure of the androgen receptor gene, which is located on chromosome Xq11-12. *Middle*, the domain structure of the receptor protein. The C-terminus contains the ligand binding domain (LBD), which is linked by hinge region to the DNA binding domain (DBD). The N-terminal domain (NTD) contains the main determinants for transactivation, termed AF1, which comprises transactivation units (TAU) 1 and 5. *Bottom* summary of the structural properties of the LBD, DBD, and NTD. The NTD is intrinsically disordered and undergoes induced folding upon protein–protein interactions; no three-dimensional structure is currently available for this region. The functional properties of the different receptor domains are summarized beneath each region

3.5.1 Ligand Binding Domain

The AR domain structure (Fig. 3.2) is shared with other members of the nuclear receptor superfamily and includes a globular ligand binding domain (LBD) in the C-terminal 250 amino acids, which consists of 11 α-helices arranged in a three layer sandwich. The AR has been shown to have similar binding affinities for both testosterone and DHT (Wilson and French 1976) and both steroids fit in an internal hydrophobic pocket, with specific binding mediated by key amino acids (N705, Q711, R752, and T877), forming both direct and water-mediated hydrogen bonding networks as well as close hydrophobic interactions with L704, M745, and F746 (Askew et al. 2007; Sack et al. 2001). The binding of hormone and activation of the AR causes a rearrangement of helix 12 and the formation of a surface on the LBD termed activation function 2 (AF2), which also involves residues from helices 3, 4, and 5. The AF2 binding pocket is involved in interactions with co-regulatory proteins and the AR-NTD (Reviewed in Kumar and McEwan 2012). The conformation of the LBD bound to testosterone or DHT is essentially the same, although testosterone dissociates about three to four-times faster (Askew et al. 2007). The reason for this difference in dissociation rates, and the potency of the two natural androgens, was correlated with local conformational differences of key residues, which link the ligand binding pocket and the AF2 surface (Askew et al. 2007). Leucine at position 712 adopts different conformations in the AR-T and AR-DHT structures and mediates interactions between methionine 745, which forms hydrophobic interactions with the steroid molecule, and amino acids making up the AF2 surface.

3.5.2 DNA Binding Domain/Hinge Region

The DNA binding domain (DBD) is a globular domain linked to the LBD by a flexible hinge of 40 amino acids. The latter contains a bipartite nuclear localization signal ([629]RKLKK[633]) that is necessary for translocation of the receptor into the nucleus. The DBD is coded for by exons 2 and 3 and represents a Cys4-type Zn-finger domain, with two ions of zinc bound by eight cysteine residues. The domain contains two α-helices that fold perpendicular to each other, which helps position the "recognition" helix into the major groove of the DNA (Fig. 3.2). In addition to mediating specific recognition and binding of DNA sequences in the genome, a five amino acid sequence ([595]ASRND[599]) acts as a dimerization interface holding two receptor monomers together on the DNA.

The AR binds to a 15 bp palindromic hormone response element (HRE), 5'-GGA/TACAnnnTGTTCT-3', as a homo-dimer. Genome-wide studies have led to the identification of networks of androgen-regulated genes and have also revealed that receptor binding sites show considerable diversity in response element architecture, including imperfect palindromic sequences and half-sites in addition to clearly recognizable consensus-like response elements (see Bolton et al. 2007; Eacker et al. 2007; Massie et al. 2007; Wang et al. 2007).

3.5.3 N-Terminal Domain

The NTD represents more than half the receptor protein (550 amino acids), and in contrast to the LBD and DBD is intrinsically disordered, and is functionally critical for receptor-dependent transactivation (Kumar and McEwan 2012; McEwan 2012). The ligand-dependent AF2 in the LBD is at best weak at activating gene expression on its own and the main determinant(s) for receptor-dependent transcriptional activation are located in the NTD and termed AF1 (Jenster et al. 1995; Simental et al. 1991). The AR-AF1 is modular in structure and maps to amino acids 142–485, which comprises both the TAU-1 (amino acids 101–360) and TAU-5 (amino acids 360–485) (Fig. 3.2) (Jenster et al. 1995).

The AR-AF1 domain has 13 % helical secondary structure as determined by biophysical analysis and four helical segments have been predicted (Kumar et al. 2004; Reid et al. 2002). Furthermore, the AR-AF1 domain has the propensity to adopt increased helical structure; in the presence of the natural osmolyte trimethylene N-oxide or the general transcription factor TFIIF binding partner there is a threefold increase in α-helix secondary structure (Kumar et al. 2004; Reid et al. 2002). Collectively the data suggests that AR-AF1 domain conforms to a "collapsed disorder" structure (Lavery and McEwan 2008) and exists as an ensemble of conformations with limited secondary and tertiary structure, which are primed to bind co-regulatory proteins. As a consequence, AR-dependent gene regulation can be thought of as resulting from the cooperation or synergism between the different receptor domains and interactions with other transcription factors (including possible tissue specific proteins) and multiple interactions with the cell's transcription machinery.

3.6 Rapid Signaling by Testosterone

As discussed earlier, the classical mechanism of AR action involves translocation into the nucleus, specific DNA binding and regulation of gene transcription (Fig. 3.3). However, there is also clear evidence for rapid effects of testosterone on target cells, which do not rely upon *de novo* transcription or protein synthesis (reviewed in Heinlein and Chang 2002; Rahman and Christian 2007). These actions occur on timescales of seconds to several minutes and are thought to involve second messenger signaling cascades, involving an influx of Ca^{2+} ions, increased cAMP levels and/or activation of different kinase pathways (Fig. 3.3).

These rapid action, sometimes referred to incorrectly as "nongenomic" may however involve the classical AR. For example, a proline repeat in the AR-NTD has been shown to mediate binding to the SH3 domain of the tyrosine kinase c-src, which leads to downstream signaling through the MAP kinase pathway (Fig. 3.3) (Cheng et al. 2007; Migliaccio et al. 2000). The classical AR has also been proposed to be tethered to the plasma memberane via palmitylation leading to

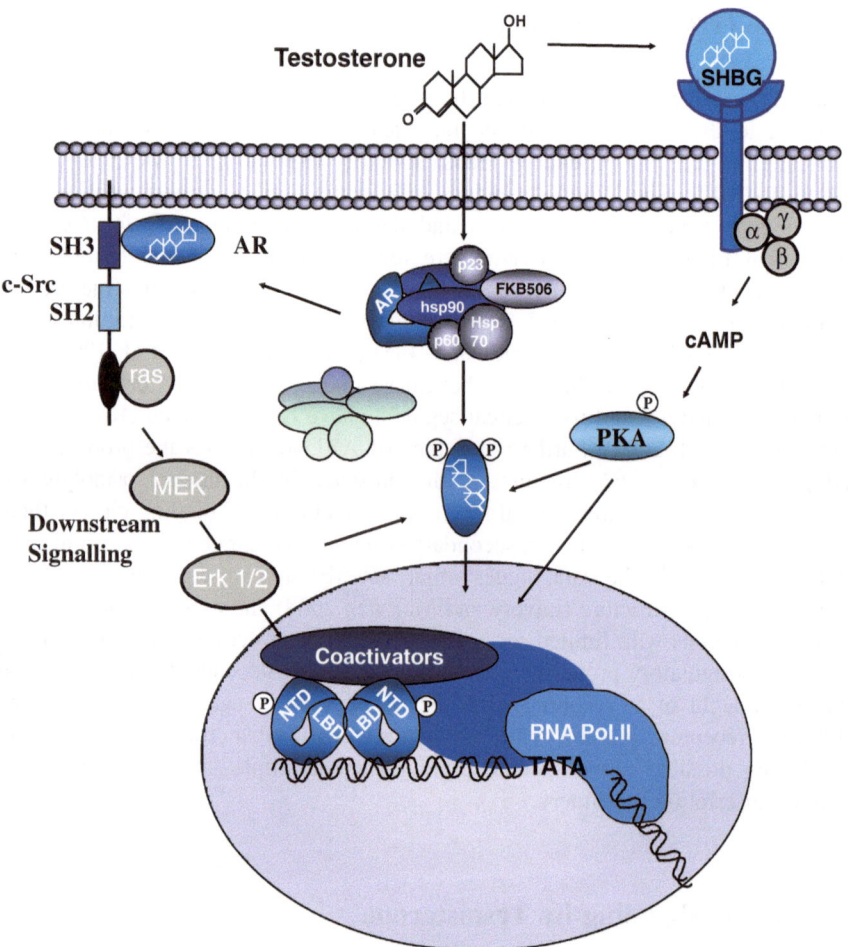

Fig. 3.3 Classical and nonclassical signaling by testosterone. The androgen receptor (AR), an intracellular protein, is bound to molecular chaperones and immunophilins (e.g., Hsp70, 90, FKB506) in the absence of hormone; upon binding testosterone the receptor translocates to the nucleus where it can bind to specific DNA elements and directly regulate gene expression (central pathway). Alternatively, in some cells, testosterone mediates rapid signaling which may be dependent or independent of the classical AR protein. For example, testosterone can interact with sex hormone binding globulin, which is bound to a cell surface receptor, and this initiates a cascade of secondary signaling molecules including cAMP (pathway on the *right*). The hormone bound AR has also be observed to interact with the tyrosine kinase c-src at the cell surface, which then signals through the MAP kinase pathway (*left* hand pathway). In both classical and nonclassical signaling, the receptor may be a substrate for post-translational modifications, such as phosphorylation (P)

downstream signaling through PI-3-kinase and MAP kinase pathways (Pedram et al. 2007). Alternatively, testosterone has been shown to work directly at the cell surface, although a membrane bound testosterone receptor remains to be formally identified. Recently, the G-coupled receptor, GPRC6A has been proposed as a candidate for a membrane testosterone receptor (Pi et al. 2010). Alternatively, rapid signaling by testosterone could result from binding to the sex hormone binding globule (SHBG), which is bound to a receptor protein on the cell surface. The SHBG receptor, although not cloned, is thought to be a G-coupled receptor which signals through adenylate cyclase to raise intracellular cAMP levels, which can then activate the protein kinase A (PKA) pathway (Fig. 3.3) (Rosner et al. 2010).

Rapid signaling by testosterone can also involve an influx of Ca^{2+}-ions, which appears to require signaling from a G-coupled receptor protein to Ca channels. Interestingly, this action also required the classical AR, as demonstrated by the use of antiandrogens (Fix et al. 2004; Gorczynska and Handelsman 1995).

What are the physiological consequences of rapid testosterone signaling? Clearly activation of different signaling cascades (i.e., MAPK, PKA) could lead to phosphorylation of many intracellular proteins. However, it is worth remembering that the classical AR is itself a target for these pathways (Fig. 3.3). Rapid actions of testosterone have been observed to be important in many cell types, including macrophages, Sertoli cells, osteoblasts (in males), myotubes, and neurons and has been correlated with gap junction communications and reorganization of the cytoskeleton in prostate cells; regulation of spermatogenesis; and neuronal cell activity (reviewed in Rahman and Christian 2007).

3.7 Key Points-Summary

- The main sites of testosterone production are Leydig cells in the testes and theca cells in the ovaries. In humans, the adrenals represent a third source of androgens.
- Testosterone acts through an intracellular receptor a protein that primarily acts as a hormone-dependent transcription factor to switch target genes on or off.
- In some cells (for example, macrophages, neuron, prostate cancer, and Sertoli cells) testosterone acts rapidly in a manner that may be independent of the classical androgen receptor.

References

Askew EB, Gampe RT Jr, Stanley TB, Faggart JL, Wilson EM (2007) Modulation of androgen receptor activation function 2 by testosterone and dihydrotestosterone. J Biol Chem 282:25801–25816
Bolton EC, So AY, Chaivorapol C, Haqq CM, Li H, Yamamoto KR (2007) Cell- and gene-specific regulation of primary target genes by the androgen receptor. Genes Dev 21:2005–2017

Chadha S, Pache TD, Huikeshoven JM, Brinkmann AO, van der Kwast TH (1994) Androgen receptor expression in human ovarian and uterine tissue of long-term androgen-treated transsexual women. Hum Pathol 25:1198–1204

Cheng J, Watkins SC, Walker WH (2007) Testosterone activates mitogen-activated protein kinase via Src kinase and the epidermal growth factor receptor in sertoli cells. Endocrinology 148:2066–2074

Eacker SM, Shima JE, Connolly CM, Sharma M, Holdcraft RW, Griswold MD, Braun RE (2007) Transcriptional profiling of androgen receptor (AR) mutants suggests instructive and permissive roles of AR signaling in germ cell development. Mol Endocr 21:895–907

Fix C, Jordan C, Cano P, Walker WH (2004) Testosterone activates mitogen-activated protein kinase and the cAMP response element binding protein transcription factor in Sertoli cells. Proc Nat Acad Sci USA 101:10919–10924

Gorczynska E, Handelsman DJ (1995) Androgens rapidly increase the cytosolic calcium concentration in Sertoli cells. Endocrinology 136:2052–2059

Heinlein CA, Chang C (2002) The roles of androgen receptors and androgen-binding proteins in nongenomic androgen actions. Mol Endcor 16:2181–2187

Hillier SG (2001) Gonadotropic control of ovarian follicular growth and development. Mol Cell Endocrinol 179:39–46

Hillier SG, Tetsuka M, Fraser HM (1997) Location and developmental regulation of androgen receptor in primate ovary. Hum Reprod 12:107–111

Hu YC, Wang PH, Yeh S, Wang RS, Xie C, Xu Q, Zhou X, Chao HT, Tsai MY, Chang C (2004) Subfertility and defective folliculogenesis in female mice lacking androgen receptor. Proc Natl Acad Sci USA 101(31):11209–11214

Jamnongjit M, Hammes SR (2006) Ovarian steroids: the good, the bad, and the signals that raise them. Cell Cycle 5:1178–1183

Jenster G, van der Korput HA, Trapman J, Brinkmann AO (1995) Identification of two transcription activation units in the N-terminal domain of the human androgen receptor. J Biol Chem 270:7341–7346

Kumar R, McEwan IJ (2012) Allosteric modulators of steroid hormone receptors: structural dynamics and gene regulation. Endocr Rev 33:271–299

Kumar R, Betney R, Li J, Thompson EB, McEwan IJ (2004) Induced alpha-helix structure in AF1 of the androgen receptor upon binding transcription factor TFIIF. Biochemistry 43:3008–3013

Lavery DN, McEwan IJ (2008) Structural characterization of the native NH2-terminal transactivation domain of the human androgen receptor: a collapsed disordered conformation underlies structural plasticity and protein-induced folding. Biochemistry 47:3360–3369

Massie CE, Adryan B, Barbosa-Morais NL, Lynch AG, Tran MG, Neal DE, Mills IG (2007) New androgen receptor genomic targets show an interaction with the ETS1 transcription factor. EMBO Rep 8:871–878

McEwan IJ (2004) Molecular mechanisms of androgen receptor-mediated gene regulation: structure-function analysis of the AF-1 domain. Endocr Relat Cancer 11:281–293

McEwan IJ (2012) Intrinsic disorder in the androgen receptor: identification, characterisation and drugability. Mol Biosyst 8:82–90

Migliaccio A;Castoria G, Di Domenico M, de Falco A, Bilancio A, Lombardi M, Barone MV, Ametrano D, Zannini MS, Abbondanza C, Auricchio F (2000) Steroid-induced androgen receptor-oestradiol receptor beta-Src complex triggers prostate cancer cell proliferation. EMBO J 19:5406–5417

Miller WL, Auchus RJ (2011) The molecular biology, biochemistry, and physiology of human steroidogenesis and its disorders. Endocr Rev 32:81–151 CVD

O'Donnell L, de Kretser DM (2013) Endocrinology of the male reproductive system. http://www.endotext.org/male/male1/male1.html. Accessed 5 June 2013

Payne AH, Hardy MP (eds) (2007) The Leydig cell in health and disease. Humana Press, Totowa

Pedram A, Razandi M, Sainson RC, Kim JK, Hughes CC, Levin ER (2007) A conserved mechanism for steroid receptor translocation to the plasma membrane. J Biol Chem 282:22278–22288

Pi M, Parrill AL, Quarles LD (2010) GPRC6A mediates the non-genomic effects of steroids. J Biol Chem 285:39953–39964

Rahman F, Christian HC (2007) Non-classical actions of testosterone: an update. Trends Endocrinol Metab 18:371–378

Reid J, Kelly SM, Watt K, Price NC, McEwan IJ (2002) Conformational analysis of the androgen receptor amino-terminal domain involved in transactivation. Influence of structure-stabilizing solutes and protein-protein interactions. J Biol Chem 277:20079–20086

Rosner W, Hryb DJ, Kahn SM, Nakhla AM, Romas NA (2010) Interactions of sex hormone-binding globulin with target cells. Mol Cell Endocrinol 316:79–85

Sack JS, Kish KF, Wang C, Attar RM, Kiefer SE, An Y, Wu GY, Scheffler JE, Salvati ME, Krystek SR Jr, Weinmann R, Einspahr HM (2001) Crystallographic structures of the ligand-binding domains of the androgen receptor and its T877A mutant complexed with the natural agonist dihydrotestosterone. Proc Natl Acad Sci USA 98(9):4904–4909

Saunders PT, Millar MR, Williams K, Macpherson S, Harkiss D, Anderson RA, Orr B, Groome NP, Scobie G, Fraser HM (2000) Differential expression of estrogen receptor-alpha and -beta and androgen receptor in the ovaries of marmosets and humans. Biol Reprod 63:1098–1105

Shiina H, Matsumoto T, Sato T, Igarashi K, Miyamoto J, Takemasa S, Sakari M, Takada I, Nakamura T, Metzger D, Chambon P, Kanno J, Yoshikawa H, Kato S (2006) Premature ovarian failure in androgen receptor-deficient mice. Proc Natl Acad Sci USA 103(1):224–229

Sidiropoulou E, Ghizzoni L, Mastorakos G (2012) Adrenal androgens. http://www.endotext.org/adrenal/adrenal3/adrenal3.htm. Accessed 5 June 2013

Simental JA, Sar M, Lane MV, French FS, Wilson EM (1991) Transcriptional activation and nuclear targeting signals of the human androgen receptor. J Biol Chem 266:510–518

Skinner MK, Griswold MD (eds) (2005) Sertoli cell biology. Elsevier, Salt Lake City

Walters KA, Allan CM, Handelsman DJ (2008) Androgen actions and the ovary. Biol Reprod 78:380–389

Wang Q, Li W, Liu XS, Carroll JS, Janne OA, Keeton EK, Chinnaiyan AM, Pienta KJ, Brown M (2007) A hierarchical network of transcription factors governs androgen receptor-dependent prostate cancer growth. Mol Cell 27:380–392

Weil SJ, Vendola K, Zhou J, Adesanya OO, Wang J, Okafor J, Bondy CA (1998) Androgen receptor gene expression in the primate ovary: cellular localization, regulation, and functional correlations. J Clin Endocrinol Metab 83:2479–2485

Wilson EM, French FS (1976) Binding properties of androgen receptors. Evidence for identical receptors in rat testis, epididymis, and prostate. J Biol Chem 251:5620–5629

Yeh S, Tsai MY, Xu Q, Mu XM, Lardy H, Huang KE, Lin H, Yeh SD, Altuwaijri S, Zhou X, Xing L, Boyce BF, Hung MC, Zhang S, Gan L, Chang C (2002) Generation and characterization of androgen receptor knockout (ARKO) mice: an in vivo model for the study of androgen functions in selective tissues. Proc Natl Acad Sci U S A 99:13498–13503

Chapter 4
Androgen Receptor Signaling in the Testis

4.1 Introduction

Testosterone is widely accepted as the main driver of spermatogenesis, being both necessary and sufficient to drive qualitative sperm development in the absence of stimulation by FSH or LH. As described above, testosterone exerts the majority of its function via binding its cognate receptor AR. AR is expressed in several cell-types of the testis through development and adult life, though importantly it is never expressed in germ cells suggesting the impact of androgens on spermatogenesis is via an indirect signaling route involving the supporting somatic cell lineages. In the mouse, AR expression begins around embryonic day 15 (Scott et al. 2007), and in week 7 of gestation in the human (Shapiro et al. 2005), and localizes to peritubular cells, which continue to express AR throughout life, and cells of the testicular interstitium. After birth AR expression is first detectable in Sertoli cells at postnatal day 4, and continues throughout life. Adult Leydig cell precursors begin to proliferate at postnatal day 12 and switch on AR expression from this point onwards. Thus in the adult testis of both rodents and humans, AR is expressed by Sertoli cells, Leydig cells, Peritubular myoid cells, vascular smooth muscle and vascular endothelial cells, all of which are responsive to androgen stimulation (reviewed in De Gendt and Verhoeven 2012a, b, c) (Fig. 4.1a).

Interestingly neither fetal Leydig cells (which produce androstenedione) nor Sertoli cells (which express 17βHSD3 and thus convert androstenedione to testosterone) (Shima et al. 2013) express AR. Thus, AR expression is not required to drive androgen production in fetal life. In essence, androgens are not required to build a testis, at least up until birth. Conversely, both AR expression in peripheral tissues and production of androgens by the fetal testis are essential for masculinization of the male fetus (see Chap. 2).

The role of AR signaling inside the testis has remained an enigma for over 100 years. While it was well established that treatment with androgens or anti-androgens influenced testis function, these treatments by their very nature were systemic, thus the entire testes was treated as an enclosed system, with all cell-types impacted, and inferences made on this basis.

L. B. Smith et al., *Testosterone: From Basic Research to Clinical Applications*,
SpringerBriefs in Reproductive Biology,
DOI: 10.1007/978-1-4614-8978-8_4, © The Author(s) 2013

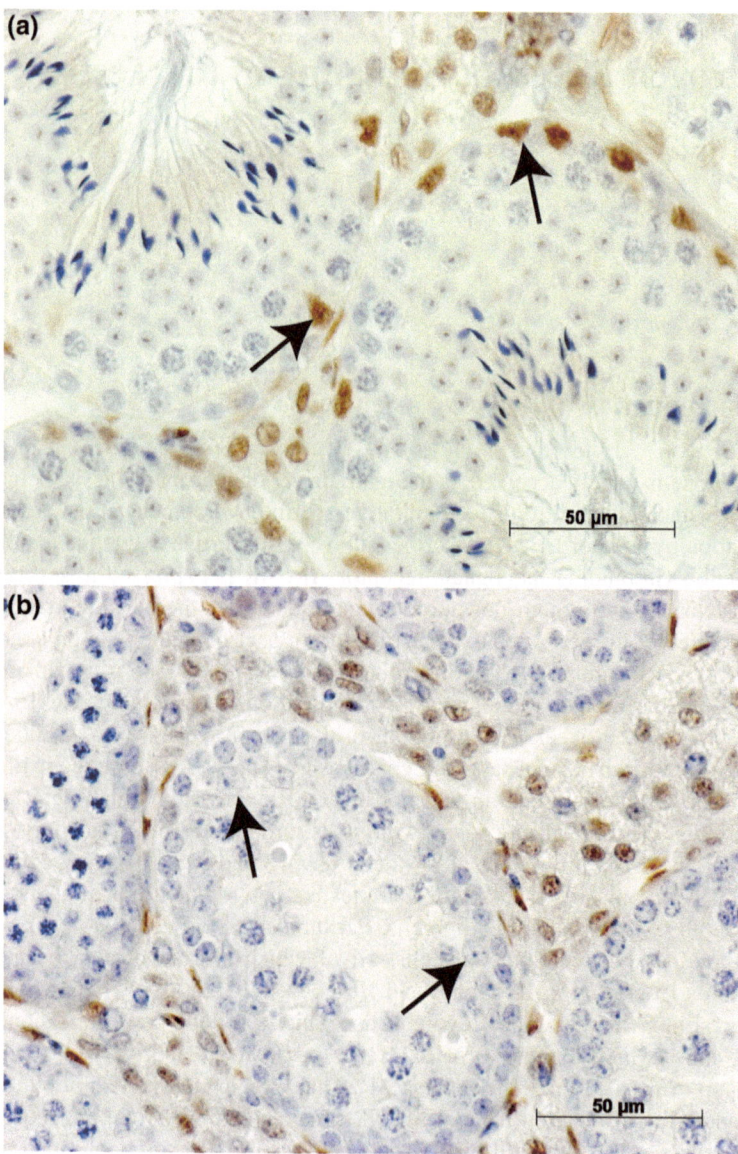

Fig. 4.1 Sertoli cell AR knockout. Androgen receptor localization in the adult mouse testis of mice and mice with Sertoli cell ARKO. **a** AR expression (*brown*) is observed in the somatic cells of the testis, including Leydig, peritubular myoid and Sertoli cell nuclei (*arrows*). **b** Cell-specific knockout of AR in the SCARKO mouse ablates Sertoli cell expression of AR (*arrows*), whilst AR expression remains unaffected in other somatic cells. Note absence of hooked-headed mature sperm in (**b**), demonstrating Sertoli cell AR is essential for maturation of germ cells

The field moved-forward significantly with the identification of the testicular feminized mouse in 1970, which possessed a Y-chromosome, yet exhibited a feminized phenotype and behavior (Lyon and Hawkes 1970). The genetic lesion in the Tfm mouse was traced to a point mutation in a nuclear receptor located on the X-chromosome, introducing a premature stop codon, which completely blocked production of the nuclear receptor protein. This protein was the AR. The Tfm mouse was used widely over the next three decades to examine the role of androgens throughout many body systems; not only masculinization, but also the cardiovascular system, diabetes, the brain, and in behavior. This model was supported in 2002 when gene-targeting was used to specifically knockout AR (the ARKO mouse) (Yeh et al. 2002). This mouse was essentially a phenocopy of the Tfm mouse, confirming the relationship between the Tfm mouse and AR.

4.2 Tfm ARKO Mouse

However, whilst incredibly important, both of these animal models had limitations in their own right. First, unlike human CAIS patients, which produce high concentrations of testosterone, these animals produce less than 10 % of the normal levels of circulating testosterone. Thus, the models not only lack AR, but also have developed in a low testosterone environment. This defect has serious confounding effects. Both DHT and estradiol are produced locally in peripheral tissues using testosterone as a substrate, these animals also have extremely low levels of DHT and estradiol. In effect this means that the Tfm/ARKO models are in fact complex models of AR ablation, low testosterone, low DHT, low estradiol (and thus low estrogen signaling). The importance of this started to come to light in studies on the vascular system where studies carried out in the ARKO mice and attributed to low testosterone, could be rescued by estradiol treatment (see Chap. 5).

A further complexity in the Tfm/ARKO mouse is the fundamental requirement of androgen-AR signaling for masculinization, and in particular the second stage of testicular descent. Thus the testes in these models were cryptorchid, that is, failed to descend into the scrotum. The effect of this is that the postnatal testis develops at too high a temperature, which has serious impacts on testis development and function. An outcome demonstrated in several studies of experimentally/surgically induced cryptorchidism (Bergh and Söder 2007). Finally, the Tfm/ARKO models fail to overcome the core problem originally presented by the systemic treatments, which is that this is once again a systemic effect. AR is ablated in all testicular cell-types bringing us no closer to dissecting the role of androgens in each cell-type of the testis. A way of squaring this circle was finally achieved in 2004, when conditional gene-targeting was exploited to knockout AR in individual testicular cell-types.

4.3 The Cre/Lox System

The development of the Cre/Lox system of conditional gene targeting, has permitted spatial and temporal localisation of genetic manipulation (for reviews, see Hadjantonakis et al. 1999; Nagy 2000). With more than 500 independent Cre Recombinase expressing mouse lines available via the CreXmice database (Nagy et al. 2009), specific deletion of a gene in just one tissue or cell-type has become a reality.

The principle underpinning the Cre/Lox system relies upon the ability of Cre Recombinase to identify, bind and recombine DNA between two *LoxP* [Locus Of crossing (X-ing) over in P1] sites, each of these 34 bp target DNA sequences consists of two 13 bp inverted repeat sequences, flanking a central, 8 bp, directional core (Smith 2011). Artificially engineering these *LoxP* sequences into DNA regions flanking a target locus of interest 'primes' that site for recombination by Cre Recombinase (Kwan 2002). Consequently, target genes/DNA sequences flanked by *LoxP* sites are said to be *'floxed'*. (Lewandoski 2001; Gossen and Bujard 2002; Branda and Dymecki 2004). The Cre lox system has been widely employed to dissect gene function in the testis (reviewed in Smith 2011).

4.4 Cell-Specific KO

To examine the cell-specific roles of AR signaling several groups have exploited the Cre-loxP systems to ablate AR in a cell-specific manner. At least five 'floxed' alleles of AR have been generated (De Gendt et al. 2004; Holdcraft and Braun 2004; Kato 2002; Notini et al. 2005; Yeh et al. 2002) (reviewed in De Gendt and Verhoeven 2012a, b, c; Walters et al. 2010).

4.5 AR in Sertoli Cells

In the mouse, AR is expressed in Sertoli cells at postnatal day 4. As such, Sertoli cell androgen receptor is not required for testis formation. Ablation of AR specifically from Sertoli cells has provided significant insights as to the role of androgens in this cell-type (De Gendt et al. 2004; Holdcraft and Braun 2004; Chang 2004; Lim 2009) (Fig. 4.1). When ablated throughout life, the adult phenotype of the testis is significant. Perhaps most surprisingly is that Sertoli cell number is not significantly reduced in the Sertoli cell knockout mouse. This suggests that the reduction in numbers of Sertoli cells in the Tfm/ARKO mouse must be the result of aberrant action in another cell-type. This was the first piece of evidence of a more complex androgen-dependent signaling system existed within the testis, as prior to this phenotypic alterations in the Tfm/ARKO mouse had always been attributed to intrinsic loss of androgen action within that cell-type. However, despite failing to impact on Sertoli cell number, these animals are completely infertile. While early stages of germ cell development continue as normal, there is a block during meiosis such that no spermatozoa are produced. This observation is consistent with the known role of Sertoli cells, supporting germ cell maturation. Furthermore, and to the surprise of everyone, when AR is ablated from Sertoli cells the number of Leydig cells present in the adult testis is reduced by around 40 % (De Gendt et al. 2005). The mechanism underpinning this observation remains to be explained, but presumably invokes the requirement for Sertoli cell-derived factors to support adult Leydig cell development.

4.6   AR in Peritubular Myoid Cells

Very little is known regarding the role of Peritubular Myoid cells in the testis. As described above, PTM cells have been ascribed a role in contractility of the seminiferous tubules, but we were unprepared for the significance of androgen signaling in this cell-type on overall testis function. In mice with cell-selective ablation of AR from PTM cells from embryonic day 17, testes are indistinguishable from littermate controls at postnatal day 12, again suggesting androgen signaling in PTM cells is dispensable for testis development (Welsh et al. 2009). Like the Sertoli cell-ARKO mice, adult PTM-ARKO males are infertile, though the cause of this is slightly different in that testes from PTM-ARKO mice show germ cells of all developmental stages, yet have significant reductions at all stages, a situation that deteriorates even further over time.

Changes in expression and localization of several key functional proteins such as Desmin and Laminin, result from the loss of AR, presumably perturbing normal PTM cell function. Loss of Peritubular Myoid AR also impacts adult Leydig cell development. While PTM-ARKO mice have normal numbers of Leydig cells in adulthood (defined as 3BHsd positive cells), Leydig cell differentiation remains incomplete in PTM-ARKO mice, with arrest in development at the progenitor stage (Welsh et al. 2009 and welsh et al. 2012)]. Interestingly, whilet PTM-ARKO mice are able to maintain circulating testosterone concentrations within the normal range, this requires increased LH input and PTM-ARKO testes are unresponsive to further stimulation by hCG. Essentially, these Leydig cells are working as hard as they can to maintain normal circulating testosterone concentrations, and rapidly moving toward failure and hypogonadism (Welsh et al. 2012). Thus in addition to their contractile role, androgen signaling in PTM cells is essential for PTM function, spermatogenesis and maturation and function of Leydig cells; further evidence of a complex androgen-dependent paracrine signaling network at work in the testis.

4.7 AR in Leydig Cells

Targeting AR in Leydig cells has been attempted, however, the choice of Cre recombinase, AmhR2-Cre proved controversial, as this Cre line targets both Sertoli and Leydig cells (Smith 2011). The resulting mutant demonstrated features that could be associated with the loss of AR in Sertoli cells, and as such are difficult to attribute to loss of AR specifically in LCs (Xu et al. 2007). Thus, the functional role of AR signaling specifically within Leydig cells remains to be established.

While attempts to describe the cell-specific roles for AR signaling in the testis continue, it is already very clear that a complex paracrine signaling network driven by androgens in different cell-types is an important regulator of testis function.

4.8 Key Points-Summary

- AR is not expressed in Germ cells, thus androgen control of spermatogenesis acts via supporting somatic cells.
- AR action is not required to develop the gross architecture of the testis, but is important for cell numbers and maturation.
- Mice with ablation of AR develop as phenotypic females with disrupted testis architecture, compounded by failure of testis descent.
- Cell-specific knockout of AR in different cells of the testis have revealed a complex network of signals that translate the androgen signal into normal testis function.

References

Bergh A, Söder O (2007) Studies of cryptorchidism in experimental animal models. Acta Paediatr 96:617–621

Branda CS, Dymecki SM (2004) Talking about a revolution: the impact of site-specific recombinases on genetic analyses in mice. Dev Cell 6:7–28

Chang C, Chen YT, Yeh SD, Xu Q, Wang RS, Guillou F, Lardy H, Yeh S (2004) Infertility with defective spermatogenesis and hypotestosteronemia in male mice lacking the androgen receptor in Sertoli cells. Proc Natl Acad Sci USA 101:6876–6881

De Gendt K, Verhoeven G (2012a) Tissue- and cell-specific functions of the androgen receptor revealed through conditional knockout models in mice. Mol Cell Endocrinol 352:13–25

De Gendt K, Verhoeven G (2012b) Tissue- and cell-specific functions of the androgen receptor revealed through conditional knockout models in mice. Mol Cell Endocrinol 352(1–2):13–25

De Gendt K, Verhoeven G (2012c) Tissue- and cell-specific functions of the androgen receptor revealed through conditional knockout models in mice. Mol Cell Endocrinol 352:13–25

De Gendt K et al (2004) A Sertoli cell-selective knockout of the androgen receptor causes spermatogenic arrest in meiosis. Proc Natl Acad Sci USA 101(5):1327–1332

De Gendt K, Atanassova N, Tan KA, de França LR, Parreira GG, McKinnell C, Sharpe RM, Saunders PT, Mason JI, Hartung S, Ivell R, Denolet E, Verhoeven G (2005) Development and function of the adult generation of Leydig cells in mice with Sertoli cell-selective or total ablation of the androgen receptor. Endocrinology 146:4117–4126

Gossen M, Bujard H (2002) Studying gene function in eukaryotes by conditional gene inactivation. Annu Rev Genet 36:153–173

Hadjantonakis AK, Pirity M, Nagy A (1999) Cre recombinase mediated alterations of the mouse genome using embryonic stem cells. Methods Mol Biol 97:101–122

Holdcraft RW, Braun RE (2004) Androgen receptor function is required in Sertoli cells for the terminal differentiation of haploid spermatids. Development 131(2):459–467

Kato S (2002) Androgen receptor structure and function from knock-out mouse. Clin Paediatr Endocrinol 11:1–7

Kwan KM (2002) Conditional alleles in mice: practical considerations for tissue-specific knockouts. Genesis 32:49–62

Lewandoski M (2001) Conditional control of gene expression in the mouse. Nat Rev Genet 2:743–755

Lim P et al (2009) Sertoli cell androgen receptor DNA binding domain is essential for the completion of spermatogenesis. Endocrinology 150(10):4755–4765

Lyon MF, Hawkes SG (1970) X-linked gene for testicular feminization in the mouse. Nature 227:1217–1219

Nagy A (2000) Cre recombinase: the universal reagent for genome tailoring. Genesis 26:99–109

Nagy A, Mar L, Watts G (2009) Creation and use of a cre recombinase transgenic database. Methods Mol Biol 530:365–378

Notini AJ et al (2005) Genomic actions of the androgen receptor are required for normal male sexual differentiation in a mouse model. J Mol Endocrinol 35(3):547–555

Scott HM, Hutchison GR, Mahood IK, Hallmark N, Welsh M, De Gendt K, Verhoeven G, O'Shaughnessy P, Sharpe RM (2007) Role of androgens in fetal testis development and dysgenesis. Endocrinology 148:2027–2036

Shapiro E, Huang H, Masch RJ, McFadden DE, Wu XR, Ostrer H (2005) Immunolocalization of androgen receptor and estrogen receptors alpha and beta in human fetal testis and epididymis. J Urol 174:1695–1698

Shima Y, Miyabayashi K, Haraguchi S, Arakawa T, Otake H, Baba T, Matsuzaki S, Shishido Y, Akiyama H, Tachibana T, Tsutsui K, Morohashi K (2013) Contribution of Leydig and Sertoli cells to testosterone production in mouse fetal testes. Mol Endocrinol 27:63–73

Smith L (2011) Good planning and serendipity: exploiting the Cre/Lox system in the testis. Reproduction 141(2):151–161

Walters KA, Simanainen U, Handelsman DJ (2010) Molecular insights into androgen actions in male and female reproductive function from androgen receptor knockout models. Hum Reprod Update 16(5):543–558

Welsh M et al (2009) Androgen action via testicular peritubular myoid cells is essential for male fertility. FASEB J 23(12):4218–4230

Welsh M et al (2012) Androgen receptor signalling in peritubular myoid cells is essential for normal differentiation and function of adult Leydig cells. Int J Androl 35(1):25–40

Xu Q, Lin HY, Yeh SD, Yu IC, Wang RS, Chen YT, Zhang C, Altuwaijri S, Chen LM, Chuang KH, Chiang HS, Yeh S, Chang C (2007) Infertility with defective spermatogenesis and steroidogenesis in male mice lacking androgen receptor in Leydig cells. Endocrine 32:96–106

Yeh S, Tsai MY, Xu Q, Mu XM, Lardy H, Huang KE, Lin H, Yeh SD, Altuwaijri S, Zhou X, Xing L, Boyce BF, Hung MC, Zhang S, Gan L, Chang C (2002) Generation and characterization of androgen receptor knockout (ARKO) mice: an in vivo model for the study of androgen functions in selective tissues. Proc Natl Acad Sci USA 99:13498–13503

Chapter 5
Androgen Signaling in Other Body Systems

5.1 Seminal Vesicles

The seminal vesicles (SVs) depend on androgen action for normal development and function (George and Wilson 1994; Mooradian et al. 1987; Wilson et al. 1981). Androgen action in SVs is mediated via DHT rather than T as SV stromal cells express 5α-reductase type 2, which converts T to DHT. This is supported by the observation that 5α-reductase knockout mice have smaller SVs (Mahendroo et al. 2001).

The SVs in adults are composed of epithelium surrounded by stromal cells, including an inner contractile layer of smooth muscle (Fig. 5.1). Adult SVs express AR in all cell types. The primary function of the SVs is to synthesize and secrete proteins as part of the seminal plasma, required for male fertility; removal of the SVs from mice impairs fertility (Peitz and Olds-Clarke 1986; Pang et al. 1979). SV secretory function is androgen dependent (Cunha and Donjacour 1987) and removal of the androgen stimulus through castration after puberty results in regression of the SVs, leading to a reduction in secretions and apoptosis of the epithelium (Deanesly and Parkes 1933); impacts that can be rescued by provision of testosterone (Deanesly and Parkes 1936), thus demonstrating the importance of androgen signaling to seminal vesicle function throughout life. In fact the plastic nature of the seminal vesicles in response to androgens means that seminal vesicle weight is often used as a biomarker of androgen action in rodent studies, with a reduction in weight signifying chronically perturbed androgen signaling.

Advances in transgenic technology have enabled cell-specific gene ablation of AR providing new opportunities to investigate the cell-specific roles for androgen action. Two models in which specific ablation of AR from the smooth muscle cells of the SVs have been generated; these mice have significantly smaller SVs in adulthood with abnormal histology and function, including loss of epithelial cell height and secretory function (Welsh et al. 2010; Simanainen et al. 2008). These results demonstrate a vital role for AR signaling via the smooth muscle cells for normal SV structure and function and reinforce the evidence that stromal-epithelial

L. B. Smith et al., *Testosterone: From Basic Research to Clinical Applications*,
SpringerBriefs in Reproductive Biology,
DOI: 10.1007/978-1-4614-8978-8_5, © The Author(s) 2013

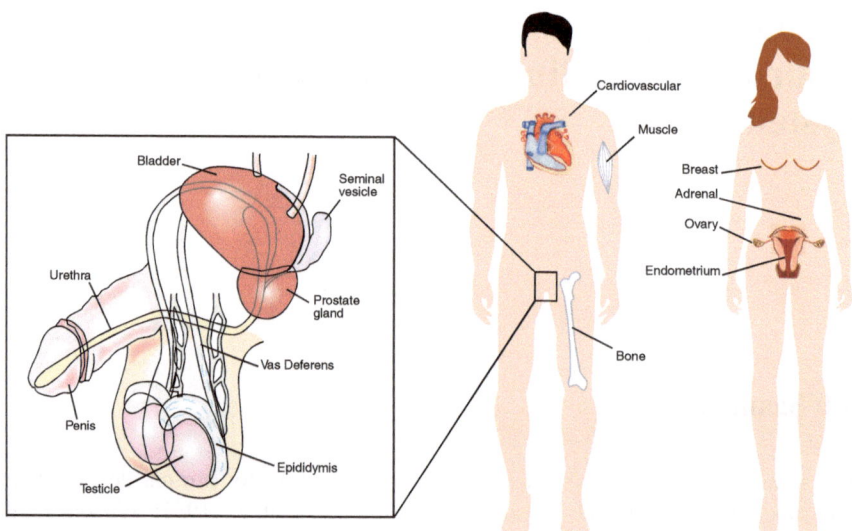

Fig. 5.1 Body-wide targets of androgens. A schematic overview of key sites of androgen action in men and women, though androgens undoubtedly play important roles elsewhere in the body. Androgen action is essential for development of the male reproductive systems and remains important for function of the prostate, seminal vesicles, penis throughout life, in addition to the essential role it plays in spermatogenesis. Outside of the reproductive system increasing evidence highlights significant roles for androgens in the cardiovascular system, muscle, and bone. In women, androgens are associated with breast, endometrium, and ovarian function

interactions are important for adult sex accessory function. However, the role of AR in seminal vesicle epithelium remains to be established.

5.2 Epididymis

The epididymis connects the efferent ducts to the vas deferens and provides a specialized environment supporting maturation and storage of sperm leaving the testis (Fig. 5.1). If sperm are prevented from completing their journey through the epididymis, the build up of sperm behind the blockage can cause obstructive azoospermia (Practice committee of ASRM 2008), a common cause of male infertility. As described in Chap. 2, both formation and function of the epididymis depend upon androgens (McPhaul 2002). Maturation and coiling of the developing Wolffian duct (the precursor to the distal end of the epididymis) are promoted by androgen action in male embryos, via binding of AR in the Wolffian duct stroma (Crocoll et al. 1998). The Wolffian duct connects to the mesonephric tubules during development (Marshall et al. 1979), which together form the epididymis, which further matures into its highly convoluted single tubule structure.

The role of testosterone in this process has been established over many years. Castration results in a decrease in epididymal weight, associated with apoptosis of epididymal epithelial principal cells (Fan and Robaire 1998; Takagi-Morishita et al. 2002) and dedifferentiation of the caput epididymal epithelium (Avram et al. 2004), which can be partially restored by the addition of testosterone (Robaire et al. 1977). Whilst castration and efferent duct ligation (which blocks fluid and sperm exiting the testis), show that both circulating testosterone and direct testicular factors control the development and function of the Initial segment (which develops from the mesonephric tubules rather than the Wolffian duct) (Fawcett and Hoffer 1979).

Genetic mutations in AR (Yeh et al. 2002; He et al. 1991) or treatment with AR antagonists to block androgen signals during the male programming window (Welsh et al. 2006) result in regression of the WD and an absence of the epididymis in adult males. Only stromal cells express AR at this time, so it has long been presumed that stromal androgen signaling is essential for epididymis development, and indeed this is the case for the Wolffian duct derived sections. The epididymal epithelium only begins to express AR just prior to birth in the mouse (embryonic day 19). Because of this late onset of expression it had long been presumed that epithelium AR played a role in epididymal function but not development.

To address this, two independent studies have recently used the Cre/lox system to ablate AR from epididymal caput epithelium (the mesonephric tubule derived structures) to establish the role of epithelial AR in the development and function of the caput epididymis (O'Hara et al. 2010; Krutskikh et al. 2010). In one of these models, AR was ablated from a proportion of proximal caput epididymal epithelial cells (O'Hara et al. 2010). In the resulting CEARKO mice, the initial segment of the epididymis completely failed to develop; an observation confirmed in the second study (Krutskikh et al. 2010). Furthermore, the cells in the remaining caput epithelium in which AR was ablated, failed to mature fully leading to a reduction in luminal diameter and impaired epithelial cell function. These abnormalities resulted in stasis of spermatozoa in the efferent ducts of CEARKO mice, which ultimately developed into a physical obstruction that resulted in fluid back-pressure, gross distension of the rete testis, and severe disruption to testicular architecture and spermatogenesis. These studies demonstrated that epithelial cell AR is a fundamental requirement for postnatal development of the IS of the epididymis; the first demonstration of a requirement for epithelial AR in the development of any androgen-dependent organ.

Together these combined data add to the sum of our knowledge, describing the critical roles androgens play in the development of the epididymis. Further research is required to establish the cell-specific functions for androgens in each cell type of the adult epididymis.

5.3 Prostate

5.3.1 Androgens and Prostate Gland Development

The prostate gland surrounds the urethra as it emerges from the bladder and in humans described as being the size and shape of a walnut (Marker et al. 2003) (Fig. 5.1). Androgenic hormones are important for both prostate development and function in adult life. The prostate contributes secreted proteins and nutrients to the seminal fluid and is therefore important for sperm viability (Marker et al. 2003). The prostate arises from the urogenital sinus at around 10–12 weeks in humans and 17.5 dpc in the mouse in response to circulating androgens. In mice and rats the branching of the prostatic ducts occurs after birth giving rise to three lobes: the anterior prostate, the dorsolateral prostate, and the ventral prostate (Marker et al. 2003). The prostate in humans although lacking a lobular morphology can be described in terms of three distinct regions: the central, transitional, and peripheral zones.

Prostatic ducts are made up of luminal secretory epithelial cells, basal epithelial cells, and stromal smooth muscle cells (Marker et al. 2003; Oldridge et al. 2012). In addition embedded within the epithelial compartment are stem cells and neuroendocrine cells (Marker et al. 2003; Oldridge et al. 2012). The importance of androgens and AR for the fetal development of the gland is clearly demonstrated by the absence of a prostate in mice and human patients lacking a functional AR (Marker et al. 2003 and references therein). Prostate development requires metabolism of testosterone to DHT by 5α−reductase type 2, which acts as the local mediator of androgen action. The AR is highly expressed in the luminal epithelial cells and is also found in the cells of the stromal compartment, but there is little or no expression in the basal cells and none in prostate stem cells (Fig. 5.2a) (Oldridge et al. 2012). *In utero* the mesenchymal AR is important for ductal morphogenesis, which involves paracrine signaling, while the epithelial AR is involved in the secretory function of the gland. Thus, both direct and indirect AR actions combine to determine prostate identity. A number of developmentally regulated genes have been identified in prostate development and key mediators of androgen action are sonic hedgehog (Shh) and the homeobox transcription factor Nkx 3.1, which are both AR-regulated (Marker et al. 2003).

5.3.2 Prostate Cancer

Prostate cancer (PCa) is a leading cause of cancer-related death in men in the developed world with an estimated 213,700 and 170,000 new cases in North America and Western Europe respectively and a around 30,000 deaths in both regions each year (Ferlay et al.2010). PCa represents an adenocarcinoma formed from the glandular epithelial cells and 60–70 % of tumors arise in the peripheral

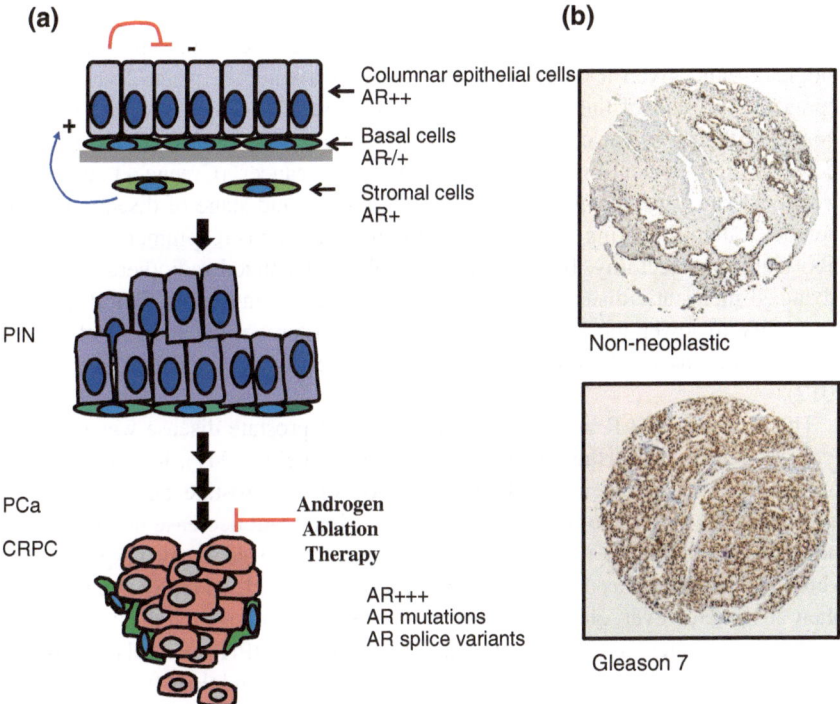

Fig. 5.2 Androgen receptor in normal and prostate cancer cells. **a** Schematic illustration showing the morphology of the normal prostate (*Top*) and the levels of androgen receptor (AR) in different cell types. The AR in stromal cells acting through paracrine factors stimulates proliferation and cell survival of epithelial cells; while the epithelial AR inhibits proliferation and promotes differentiation and regulates protein synthesis and secretion. Prostate cancer development involves multiple genetic changes, indicated by the *arrows*, including the loss of tumor suppressor gens NKx3.1, PTEN, and Rb; the early stages are thought to be characterized by development of prostatic intraepithelial neoplasia (PIN). Treatment of metastatic disease involves blocking testosterone production and AR signaling (androgen ablation therapy). However, resistance to therapy results in castrate-resistant prostate cancer (CRPC), which is associated with overexpression of the AR protein, point mutations in the receptor protein, and/or the presence of splice variants lacking the AR-LBD. **b** Immunohistochemical staining of the AR protein in high Gleason score tumor and matched non-neoplastic tissues. In the cancer tissue there is increased levels of the receptor protein and loss of the glandular structure (Data provided by Dagmara McGuinness and IJM)

zone with 25 % occurring in the transitional zone. Age and race, along with family history, are commonly cited as the major risk factors. PCa can arise at multiple foci in the same prostate and can demonstrate genetic heterogeneity (Greene et al. 2008; Oldridge et al 2012). High grade prostatic intraepithelial neoplasia (PIN), which is associated with genetic changes in luminal-like cells, is thought to represent a precursor stage for developing PCa (Shen and Abate-Shen 2010; Oldridge et al. 2012). The aggressiveness of the tumor is an indicator of whether the cancer is organ confined or has spread beyond the capsule, with bone and lymph nodes being common sites of metastasis (Fig. 5.2a). Tumors are commonly graded histologically and given a Gleason Score, which is a value between 2 and 10 and represents the sum of the individual scores for two foci. Tumors with a Gleason score less than 5 is not considered aggressive, with 5–7 considered low risk and scores 8–10 representing high grade, aggressive cancer (Greene et al. 2008). Treatment can vary depending on age of the patient and stage of disease and can include: watchful waiting, where no specific intervention is recommended, surgery involving prostatectomy, brachytherapy involving localized radiotherapy and for advanced metastatic disease androgen ablation therapy involving GnRH agonists or surgery to lower the levels of circulating testosterone and the use of antiandrogens, such as bicalutamide (Casodex), to block AR activity (Oldridge et al. 2012).

The blocking of AR action for the treatment of prostate disease was proposed over 70 years ago by Huggins and co-workers (Huggins 1941). Removing circulating levels of androgens result in involution of the prostate and apoptosis of epithelial cells and cancer cells. There is also a significant decrease in the levels of the androgen-regulated prostate specific antigen (PSA) gene; overexpression of which is widely used as a biomarker for PCa (Greene et al. 2008; Shen and Abate-Shen 2010). However, elevated levels of PSA is not necessarily indicative of PCa, as conditions such as benign prostatic hyperplasia (BPH) will also cause an increase in PSA, and cannot unambiguously distinguish indolent from aggressive tumors. Androgen ablation treatment typically looses its effectiveness after 18–36 months, due to hormone therapy resistance of the tumor cells leading to castrate-resistant prostate cancer (CRPC) (Lamont and Tindall 2011; Knudsen and Kelly 2011). Resistance to hormone therapy is associated with amplification of the AR gene and increased expression of the receptor protein (Fig. 5.2b); mutations in the AR coding sequence; expression of splice variants lacking the ligand binding domain; cross-talk with growth factor signaling pathways; altered expression of receptor coregulatory proteins; and intratumor androgen biosynthesis (Shen and Abate-Shen 2010; Dehm and Tindall 2011; Knudesen and Kelly 2011; Lamont and Tindall 2011; Hay and McEwan 2012; Waltering et al. 2012). Given the heterogeneous nature of prostate tumors it is likely more than one mechanism is involved, but crucially it is recognized that the AR remains functional and continues to be an important driver of cancer progression and that the receptor remains a valid drug target in CRPC (Chen et al. 2008; Knudsen and Kelly 2011; Lamont 2011). Recent advances in targeting the AR in CRPC have involved in developing new antiandrogens, such as enzalutamide (Tran et al. 2009); identifying novel receptor inhibitors that target the AR-NTD (Andersen et al. 2010); and a CYP17 inhibitor, abiraterone, that blocks the tumor cells ability to synthesis testosterone (Attard et al. 2008).

Androgens play a dual role in maintaining prostate cell differentiation and at the same time promote cell proliferation and survival, which as discussed above is likely to involve a mesenchymal-epithelial cell interaction (Shen and Abate Shen 2010). Thus, different functions have been assigned to the AR present in stromal and epithelial cells; through inducing paracrine factors (i.e., FGF10) the stromal AR stimulates epithelial cell proliferation and survival. Conversely, the epithelial cell AR inhibits cell proliferation and promotes cell differentiation (Fig. 5.2a). Evidence supporting the complex interplay between the stromal and epithelial compartments, both in normal development of the prostate and, in the initiation and progression of PCa has come from cell culture studies and mouse models. Loss of the epithelial AR blocks paracrine induced PIN (Memarzadeh et al. 2011) and conversely knocking down the epithelial receptor increases cell invasiveness, apoptosis of luminal epithelial cells and proliferation of basal cells (Wu et al. 2007; Niu et al. 2008a). Collectively these findings support a role for the epithelial AR in tumor promotion and as a tumor suppressor depending on the stage of the disease. This is further demonstrated by the knockout of the stromal AR, which blocks early stage tumor proliferation (Niu et al. 2008b), while ablation of the receptor in smooth muscle cells leads to a smaller prostate that is histologically abnormal (Welsh et al. 2011). Furthermore, a loss of function mutation in the AR results in impaired receptor-dependent growth inhibition (D'Antonio 2010). Collectively, the evidence from cell-selective AR knockout mice supports a role for the receptor in both PCa initiation and progression. In addition to the role for both androgens and AR in PCa progression a number of other key genetic changes have been correlated both with the early stages and advanced disease. Notably the loss of the tumor suppressors NKx 3.1 and PTEN are involved in PCa initiation, while the loss of the cell-cycle regulator Retinoblastoma protein (Rb) has been correlated with advanced and metastatic disease associated with the castrate-resistant phenotype (CRPC) (Greene et al. 2008; Macleod 2010 and references therein).

5.4 Muscle and Bone

As discussed earlier (Sect. 5.3) inhibiting AR activity is of benefit in the management of advanced prostate cancer, but there are also disadvantages as anabolic effects of androgens will also be impaired. Conversely, there are conditions such as hypogonadism and possibly aging, where it is beneficial to give androgens (see Chap. 6). However, the growth promoting action of androgens on the prostate is of real concern with androgen replacement therapies. One solution to this dilemma has been to identify synthetic receptor ligands which would promote anabolic AR responses, by maintaining muscle mass and bone integrity, but which fail to activate the receptor and growth in prostate cells (Schulman 2012; Wilson 2007 and references therein). Such compounds have been termed 'selective androgen receptor modulators' or SARMs and bind to the AR-LBD in place of testosterone (Narayanan et al.2008; Haendler and Cleve 2012; Bhasin 2009). To date a number of non-steroidal molecules have been designed and pre-clinical studies have provided encouraging results and a number of these candidate SARMs are in clinical trials for indications such as sarcopenia and osteoporosis (Narayanan 2008; Haendler 2012).

However, unraveling the molecular mechanisms of testosterone action in bone and muscle tissues remains poorly understood and is an important area of pre-clinical and clinical research. Understanding how testosterone acts in these tissues is essential if we hope to realize the potential therapeutic benefits of testosterone replacement in chronic diseases ranging from HIV wasting to cancer cachexia as well as frailty associated with aging, without adverse proliferative actions in the prostate.

5.4.1 Bone

While it is reasonable to consider the skeleton as a stable ridged structure that provides support and strength to our bodies it is also a dynamic system that involves growth, repair, and remodeling (Fig. 5.1). The two principal cells of bone are the bone forming osteoblasts and the bone absorbing osteoclasts, both of which express the AR, and both are found in trabecular and cortical bone. The actions of testosterone in bone physiology are mediated both directly through the AR and indirectly through conversion to estradiol by the enzyme aromatase (Notelovitz 2002; Callewaert et al. 2010). A osteoblast-specific knockout of the AR results in increased bone reabsorption, decreased structural integrity, and a reduction in trabecular bone mass in male animals (Notini et al. 2007). While knocking out the AR in terminally differentiated osteoblasts revealed an overlapping phenotype with the earlier study, also showed the receptor was important for mineralization and the architecture of both trabecular and cortical bone (Chiang et al. 2009). These data, together with the findings from the complete AR knockout animals highlight a role for the AR in maintaining bone mass and mineral density (Kawano et al. 2003; Imai et al. 2009; Sinnesael et al. 2011; De Gendt and Verhoeven 2012).

Animal studies suggest that the AR is not required for longitudinal or cortical radial bone growth in male mice, but was necessary for maintaining trabecular bone mass (reviewed in Sinnesael et al. 2011). In human studies there is a strong correlation in men between estrogen levels and bone physiology, but at the same time generally concluding that androgens play an important role in bone formation and maintenance. This is particularly clear in hypogonadal men or men undergoing androgen ablation therapy for prostate cancer, where testosterone administration prevents bone reabsorption and maintains bone mass (Sinnesael et al. 2011).

5.4.2 Muscle Cells

Testosterone is well known to have anabolic effects in skeletal muscle, increasing protein synthesis and causing muscle hypertrophy (reviewed in Du Bois et al. 2012) (Fig. 5.1). However, as in bone (above), the effects of testosterone are mediated both directly through the AR and indirectly through conversion to estrogens or cross-talk with other growth factor signaling pathways (i.e., IGF-I). Expression of the AR is observed in a number of muscle cell types including satellite cells, fibroblasts, progenitor cells, and myocytes (Sinha-Hikim et al. 2004). Furthermore, genome-wide chromatin binding studies and transcription analysis have identified receptor binding sites in myoblasts and a number of directly regulated AR-target genes including Myo D, Mef2c, Odc1 (Wyce et al. 2010).

The physiological consequences of androgens are further complicated by the observation that different muscle fibers respond differential to testosterone. Thus, levator ani (LA) and bulbocavernous (BC) skeletal muscle require androgens for maintenance and activity, in contrast to the limb skeletal muscle extensor digitorum longus (EDL) (Du Bois et al. 2012). Two myocyte selective AR-knock models further illustrate the differential actions of the AR in different muscle fibers. In a post-mitotic myocyte-specific AR knockout there is a significant decrease in protein expression and muscle weight in both LA and EDL, but not other hind limb muscles. In the second model (using a different *Cre*-line) there is again a decrease in LA weigh, but not in the hind limbs studied, although in contrast to the earlier model a decrease in hind limb muscle strength was observed (reviewed in MacLean and Handelsman 2009; De Gendt and Veroeven 2012).

5.4.3 Spinal Bulbular Muscular Dystrophy (SBMA; Kennedy's Disease)

The protein sequence of the AR demonstrates striking polymorphisms in two large amino acid repeats located in the NTD: representing stretches of poly glutamine (Q) and poly-glycine residues, respectively (Choong & Wilson. 1998; McEwan 2001; Palazzolo et al. 2008). The normal poly-Q repeat varies between 9 and 36, while a repeat of 38 to 62 residues is the pathological cause of spinal bulbar muscular atrophy (SBMA) or Kennedy's disease (LaSpada et al. 1991; Poletti 2004; Tanaka et al. 2012). SBMA is a late onset sex-limited neuromuscular degenerative disease associated with motor neuron cell-death in the brain stem and spinal cord leading to progressive muscle wasting. Despite extensive research using both cell culture and in vivo animal models a number of question regarding the pathology of the disease remain open to interpretation (Poletti 2004; Monks et al. 2008; Jordan and Lieberman 2008; Tanaka et al. 2012).

A wide range of experimental evidence, including the absence of SBMA in patients with AIS (see Chaps. 2 and 6), support a model for the pathology involving a 'toxic gain of function' due to expansion of the poly-Q repeat. Studies from animal models and cell culture have suggested that hormone-activation, aggregation, fragmentation, and/or subcellular localization of the AR could all play a role in disease etiology or progression. However, the mutant AR is known to be transcriptionally less active and shows reduced expression levels, therefore although less likely, it could be argued that a loss of normal receptor function may be a contributing factor to the disease phenotype (Monks et al. 2008; Tanka et al. 2012 and references therein).

Traditionally, it has been thought that the presence of the mutant AR in motor neurons resulted in cell death and this in turn leads to muscular degeneration. However, more recent studies suggest that neuronal cell dysfunction maybe the critical event. Furthermore, an animal model where the wild-type AR has been overexpressed specifically in skeletal muscle cells resulted in a phenotype similar to SBMA (Monks et al. 2008), raising the question as to what is the primary cell type responsible for the disease: the motor neuron or the skeletal muscle? (Jordan and Lieberman 2008; Monks et al. 2008).

Similarly, questions remain regarding the role of protein aggregation: Formation of inclusions in both the cytoplasm and the nucleus have been observed in motor neuron cells in culture, in vivo models and autopsy samples and is generally accepted as a defining feature of SBMA. However, what has been debated in the literature is whether these are part of a causal or protective mechanism or perhaps just a consequence of deregulation of the receptor function (reviewed in McEwan 2001; Monks et al. 2008). One solution to this problem might be that large insoluble protein aggregates represent the cell's attempt to process and remove the mutant AR, while soluble microscopic aggregates represent the pathological agent.

A more consistent picture emerges regarding the role of testosterone and nuclear translocation of the AR. Testosterone treatment generally causes an increase in receptor aggregates and is recognized as a trigger for the disease due to activation of the mutant AR and translocation to the nucleus. Thus, in animal models surgical or chemical castration, using a LHRH analog, can rescue animals from disease (Jordan and Lieberman 2008; Tanaka et al. 2012). However, translation of these findings to human clinical trials and possible therapies has only met with limited success to date (Tanaka et al. 2012).

The actions of androgens in bone and muscle tissues are clearly complex and multifactorial, involving both direct (AR mediated) and indirect mechanisms (conversion to estradiol). However, the increasing number of animal models together with identification of target genes is helping to dissect the anabolic functions of androgenic steroids.

5.5 Cardiovascular Role of Testosterone

Cardiovascular disease (CVD) is the single biggest cause of death and disability in the developed world, accounting for approximately 40 % of all deaths annually. Consequently, prevention of coronary heart disease is one of the most pressing health requirements in the developed world. Improving our understanding of the mechanisms that contribute to the development of cardiovascular disease is of primary importance in developing strategies to combat this condition.

Sex hormones have been proposed as important modulators of cardiovascular risk as men are twice as likely as women to die from coronary heart disease

(Wingard et al. 1983; Regitz-Zagrosek 2006) and cardiovascular risk increases dramatically in women post-menopause. However, the part played by sex hormones in these observed gender differences in CVD risk is complex and far from clear. Initial evidence from large observational studies that hormone replacement therapy (HRT) was cardioprotective in women has been challenged by a large randomized controlled trial (in over 16,000 women) which suggested that HRT increased risk of cardiovascular death and non-fatal myocardial infarction (Rossouw et al. 2002). Following this landmark trial, the use of HRT in women fell dramatically around the world and research interest shifted to the possibility that androgens play a key role in the risk and pathophysiology of CVD (for review see Kaushik et al. 2010) (Fig. 5.1). The increased risk of severe cardiac hypertrophy and sudden death that can result from abuse of anabolic steroids (Sullivan et al. 1998) does not reflect the wider evidence from animal and human research indicating that androgens may in fact be beneficial in cardiovascular disease (Naghi et al. 2011). First, there is strong epidemiological evidence for a protective role of androgens in CVD. The prospective EPIC Norfolk study followed 11 606 men for 6–10 years and found an inverse relationship between endogenous testosterone levels and overall mortality and cardiovascular disease (Khaw et al. 2007) (low testosterone levels are associated with increased prevalence of cardiovascular risk factors, including: hypertension, diabetes, obesity, and a prothrombotic state) (English et al. 1997). Second, testosterone can modify the clinical manifestations of disease; for example, in patients with hypogonadism and angina, androgen supplementation results in a higher threshold for myocardial ischaemia (Rosano et al. 1999). Third, testosterone administration reduces atherosclerotic plaque formation in cholesterol-fed rabbits and mice (Alexandersen et al. 1999; Nathan et al. 2001). Androgens may also play a key role in the response of the vascular wall to injury. Neointimal proliferation following coronary artery balloon angioplasty is more marked in the absence of endogenous testosterone in swine (Tharp et al. 2009) and testosterone inhibits coronary artery smooth muscle cell proliferative-phenotype by blocking cell cycle progression (Bowles et al. 2007). Clinical studies comparing men and women are more varied, with some suggesting an increased risk of restenosis following bare metal coronary stent insertion in women and some suggesting no difference (Presbitero et al. 2008; Set et al. 2010). While neointimal proliferation as a process is quite distinct from atherosclerosis it is widely accepted to be a marker of underlying vascular responsiveness to injury which itself is one component of the atherogenic process.

Together these data correlating reduced androgen action to increased risk of CVD is both a significant health and economic concern in increasingly aging Western societies. It is now well established that blood testosterone concentrations reduce progressively as men age (Travison et al. 2007; Wu et al. 2008) potentially reducing their protective role, and there has been a well-documented reduction in average blood testosterone levels in Western men over the past 40 years (Andersson et al. 2007; Bhasin 2007; Travison et al. 2009), suggesting any reduction in cardioprotection will be more prevalent over the coming decades.

Thus, characterization of how androgens impact on cardiovascular pathophysiology is extremely timely, although remains extremely challenging to dissect.

One reason why the influence of androgens on cardiovascular pathophysiology is difficult to determine is that the mechanisms through which these hormones influence cardiovascular physiology are not fully understood. Circulating concentrations of androgens rise steeply in males during puberty and are approximately eight to ten-fold higher in men than in women of reproductive age. Androgens affect cellular function by both non-genomic (fast-acting) mechanisms and by signaling through the AR (see Chap. 3), to modulate transcription of specific target genes. However, whilst vascular endothelial cells (ECs) and vascular smooth muscle cells (SMCs) both express androgen receptors (Liu et al. 2003), the role of these receptors in modulating vascular effects of androgens in health and disease remains unknown. Testosterone has been shown to induce direct, endothelium-independent vasodilatation in a variety of vascular territories but this response appears to be mediated by a non-genomic pathway, independent of AR; possibly by receptors in the cell membrane (Jones et al. 2003). The physiological significance of this response is questionable, however, since it requires supra-physiological concentrations of testosterone in arteries analyzed ex vivo. The impact of exposure to testosterone on vascular function is also unclear; long-term testosterone administration (to female monkeys or spontaneously hypertensive rats) increased endothelium-dependent relaxation (Adams et al. 1995; Tatchum-Talon et al. 2002). In contrast, acute exposure to testosterone inhibits endothelium-dependent relaxation (Ceballos et al. 1999; Teoh et al. 2000). A further complication is that, like in other tissues such as bone and muscle, testosterone is readily converted to 17β-estradiol by the enzyme aromatase and, therefore, vasodilator effects of testosterone reported in vivo may in fact be due to the activation of estrogen receptors (ER). The complex mechanisms of androgen action on the vascular system have made it extremely challenging to characterize the significance or impact of androgen signaling specifically via AR on cardiovascular function. However, intriguingly, several previous studies have correlated androgen signaling via AR with an important cardioprotective role. For example, in men without clinically manifest coronary heart disease, a reduction in AR expression in the media of coronary arteries correlates with an increase in atherosclerotic plaque area (Liu et al. 2005). This is further supported by in vitro experiments on cultured rabbit aortas in which testosterone-mediated inhibition of plaque formation was associated with greater AR expression (Hanke et al. 2001). The ability for AR signaling to modulate cardiac function has also been suggested in a study demonstrating that testosterone (or its metabolite DHT) induces hypertrophy in cultured cardiomyocytes, and that this response could be abolished by AR-antagonism (Marsh et al. 1998).

A recent resurgence in the concept of AR as a specific mediator of vascular health has followed from advances in understanding provided by cell-signaling studies. Testosterone was recently shown to act in a cardioprotective manner in cultured SMCs to inhibit vascular calcification via AR-dependent transactivation of Gas6, a key regulator of inorganic phosphate-induced calcification of vascular smooth muscle cells (Son et al. 2010). Furthermore, AR protects against angiotensin II-induced vascular remodeling via the preservation of NO bioavailability. The mechanism for this is activation of the Akt-eNOS system which suppresses remodeling via regulating oxidative stress, c-Jun, JNK signaling, and the TGF-beta-phospho-smad pathway (Ikeda et al. 2009). A further study demonstrated that androgens, acting via AR, stimulate Human Aortic Endothelial Cell (HAoEC) proliferation through up-regulation of VEGF-A, cyclin A, and cyclin B, which the authors suggest may be beneficial to cardiovascular functions as endothelial cell proliferation could assist the repair of endothelial injury/damage in the cardiovascular system (Cai et al. 2011). Perhaps most intriguingly, testosterone has recently been shown to stimulate an AR-dependent, but non-genomic, activation of eNOS. Blocking with an AR-specific antagonist (nilutamide), or AR-specific siRNAs, abolished this effect (Yu et al. 2010). Together these data implicate AR activation as an important regulator of cellular processes relevant to cardiovascular disease and remodeling. Investigation of these hypotheses in a valid and tractable animal model system is therefore extremely timely.

The testicular feminized mouse (*Tfm*) model, which exhibits complete androgen receptor ablation resulting in a model of the human condition "complete androgen insensitivity syndrome," has been employed in an attempt to dissect the specific function genomic androgen signaling (via AR) plays in vascular function. Use of isolated arteries from this model (Jones et al. 2003) has demonstrated reduced endothelial cell function and a potential pre-disposition to cardiovascular disease; although acute testosterone-mediated relaxation in vitro was unaltered (supporting the conclusion that this response is not AR-dependent). More recently, it has been suggested that enhanced atherosclerotic lesion development in *Tfm* mice is inhibited by testosterone replacement; an effect suggested to be independent of AR but possibly mediated in part by conversion of testosterone to 17β-estradiol (Nettleship et al. 2007). Thus, despite detailed analysis of the *Tfm* model, a role for the AR expressed widely in the cells of the cardiovascular system remains undefined. Perhaps one reason this information has not been forthcoming, is that the *Tfm* mouse has a major limitation for use as a model of vascular AR function. Namely, the *Tfm* mouse exhibits markedly reduced circulating concentrations of testosterone (<10 % serum testosterone levels of a normal male), as a result of 17β-hydroxylase deficiency in the testicular Leydig cells (Murphy and O'Shaughnessy 1991). Consequently, the vasculature in *Tfm* mice has not developed with exposure to normal levels of androgens and is, therefore, a model of both AR ablation *and* reduced non-genomic androgen action. Because of this, a further limitation of the *Tfm* mouse is that any studies performed in vivo require pharmacological testosterone replacement. Another limitation of the *Tfm* mouse is that it does not allow discrimination between the effects elicited by AR in different cellular locations (see Chap. 4). In recent studies testosterone reduced lesion formation in apolipoprotein E knockout mice, whilst loss of AR function (*Tfm*) in the same atherosclerosis-prone mice increased lesion size (Bourghardt et al. 2010). Furthermore, the androgen-AR system protected against Doxirubicin-induced cardiotoxicity (Ikeda et al. 2010). However, the body-wide ablation of AR in the

Tfm model prevents characterization of the cell-specific role of AR in such cardiovascular pathologies. A more elegant approach to studying the role of androgen-mediated AR stimulation in regulation of cardiovascular physiology and pathophysiology in vivo than the *Tfm* mouse, is the use of transgenic techniques to produce mice with *cell-selective* AR deletion in which circulating levels of testosterone in males and females remain unaffected.

Direct evidence for the role of AR in specific vascular cell types has come from such vascular cell-specific AR ablation models. Ironically perhaps, the first published studies have examined the role of AR in the vascular system of the testis. Whereas AR expression in vascular endothelial cells (VEARKO mice) appears to have minimal impact on vascular function (O'Hara and Smith 2012), cell-specific ablation of AR from vascular smooth muscle (SMARKO mice) leads to a significant increase in interstitial fluid volume within the testis, due to impairment of testicular vasomotion (Welsh et al. 2011) (the androgen-dependent rhythmical contraction and relaxation of the arterioles). Thus, whilst these studies remain at a preliminary stage, androgens would appear to exert at least part of their influence via the smooth muscle cells, by controlling contractile response.

5.6 Female Reproductive System

In female reproductive tissues the AR is involved in complementing and/or antagonizing the actions of other sex hormones, for example estrogens in breast and estrogens and progesterone in the uterus. The main circulating androgens in woman, produced by the ovaries and adrenal glands, are testosterone, dehydro-epiandrosterone (DHEA) and androstenedione: and in both pre- and post-menopausal woman the ovaries and adrenals contribute equally to serum testosterone levels (2–3 nM) (Labrie et al. 2003). Although the levels of adrenal DHEA fall dramatically after the age of 30 years, in both men and woman, it remains important physiologically as it can be converted to testosterone in peripheral tissues through the enzyme 17β-hydroxysteroid dehydrogenase (HSD). The contribution of peripheral tissues to the biosynthesis of active steroids, such as testosterone and estradiol, is referred to as intracrinology (reviewed in Labrie et al. 2003).

5.6.1 Breast

In normal breast tissue (Fig. 5.1) the AR displays a growth inhibitor function and antagonizes the action of the estrogen receptor (ER) α(NR3A1). The AR is expressed in the luminal epithelial cells, stromal fibroblasts, and adipocytes (Hickey et al. 2012). The AR is also prevalent in breast cancer and there is debate in the literature as to its value as a prognostic marker and as a therapeutic target. The available data suggests that in ERα + tumors, an increase in the levels of the

AR maybe an attempt to compensate for an imbalance in ERα levels or activity and that the failure of androgen antagonism leads to a growth advantage and disease progression. Conversely in tumors which are ERα -/HER2+ the AR may act as an oncogene (Labrie et al. 2003; Hickey et al. 2012). Interestingly, AR upregulated the tumor suppressor PTEN in breast cancer, whilst in prostate cancer the receptor had the opposite action, inhibiting PTEN expression. The consequence of this differential regulation was that in breast cancer increased AR and PTEN represented a better chance of disease free survival, while the upregulation of the AR and downregulation of PTEN in prostate cancer indicated a poor clinical outcome (Wang et al. 2011).

5.6.2 Endometrium

The AR is found in endometrial stromal cells (Fig. 5.1) and the levels change during the menstrual cycle, with the highest expression seen during the proliferative phase, falling during secretory phase and little or no receptor protein detected in the late secretory phase (reviewed in Cloke and Christian 2012). Progesterone regulates the differentiation of stromal cells into decedual cells, which produces secretory cells and prepares the womb for possible embryo implantation. However, from clinical studies of infertility in conditions such as premature ovarian failure (POF) and polycystic ovarian syndrome (PCOS) (see Chap. 6), associated with depletion and excess androgens respectively, together with receptor knockout animal models it is now known that androgens play an important role in female fertility and endometrial physiology (reviewed in Walters et al. 2008; Cloke et al. 2012; and references therein).

The role of testosterone and the AR in female reproductive physiology is attracting increased research interest. The importance of the receptor in normal physiology and fertility, as well as hormone-dependent cancer, clearly requires further investigation; and recent work has focused on identifying androgen-regulated genes and expression of the receptor in different cell types (see Cloke et al. 2008; McEwan et al. 2010; Marshall et al. 2011).

5.7 Key Points-Summary

- Testosterone has profound actions in both male and female reproductive tissues and impacts upon fertility and pathology.
- The AR is also widely expressed in non-reproductive tissues, including the cardiovascular system, bone, and muscle.
- The actions of testosterone are complex and multifactorial involving the AR and conversion to 17β-estradiol and cross talk with growth factor signaling (i.e. TGFβ, IGF-1).

References

Adams MR, Williams JK, Kaplan JR (1995) Effects of androgens on coronary artery atherosclerosis and atherosclerosis-related impairment of vascular responsiveness. Arterioscler Thromb Vasc Biol 15:562–570

Alexandersen P, Haarbo J, Byrjalsen I, Lawaetz H, Christiansen C (1999) Natural androgens inhibit male atherosclerosis: a study in castrated, cholesterol-fed rabbits. Circ Res 84:813–819

Andersen RJ, Mawji NR, Wang J, Wang G, Haile S, Myung JK, Watt K, Tam T, Yang YC, Banuelos CA, Williams DE, McEwan IJ, Wang Y, Sadar MD (2010) Regression of castrate-recurrent prostate cancer by a small-molecule inhibitor of the amino-terminus domain of the androgen receptor. Cancer Cell 17:535–546

Andersson AM, Jensen TK, Juul A, Petersen JH, Jorgensen T et al (2007) Secular decline in male testosterone and sex hormone binding globulin serum levels in Danish population surveys. J Clin Endocrinol Metab 92:4696–4705

Attard G, Reid AH, Yap TA, Raynaud F, Dowsett M, Settatree S, Barrett M, Parker C, Martins V, Folkerd E, Clark J, Cooper CS, Kaye SB, Dearnaley D, Lee G, de Bono JS (2008) Phase I clinical trial of a selective inhibitor of CYP17, abiraterone acetate, confirms that castration-resistant prostate cancer commonly remains hormone driven. J Clin Oncol 26:4563–4571

Avram C, Yeung CH, Nieschlag E, Cooper TG (2004) Regulation of the initial segment of the murine epididymis by dihydrotestosterone and testicular exocrine secretions studied by expression of specific proteins and gene expression. Cell Tissue Res 317:13–22

Bhasin S (2007) Secular decline in male reproductive function: Is manliness threatened? J Clin Endocrinol Metab 92:44–45

Bhasin S, Jasuja R (2009) Selective androgen receptor modulators as function promoting therapies. Curr Opin Clin Nutr Metab Care 12:232–240

Bourghardt J, Wilhelmson AS, Alexanderson C, De Gendt K, Verhoeven G et al (2010) Androgen receptor-dependent and independent atheroprotection by testosterone in male mice. Endocrinology 151:5428–5437

Bowles DK, Maddali KK, Dhulipala VC, Korzick DH (2007) PKCdelta mediates anti-proliferative, pro-apoptic effects of testosterone on coronary smooth muscle. Am J Physiol Cell Physiol 293:C805–C813

Callewaert F, Boonen S, Vanderschueren D (2010) Sex steroids and the male skeleton: a tale of two hormones. Trends Endocrinol Metab 21:89–95

Cai J, Hong Y, Weng C, Tan C, Imperato-McGinley JL et al (2011) Androgen Stimulates Endothelial Cell Proliferation via an Androgen Receptor-VEGF/Cyclin A Mediated Mechanism. Am J Physiol Heart Circ Physiol

Ceballos G, Figueroa L, Rubio I, Gallo G, Garcia A et al (1999) Acute and nongenomic effects of testosterone on isolated and perfused rat heart. J Cardiovasc Pharmacol 33:691–697

Chen Y, Sawyers CL, Scher HI (2008) Targeting the androgen receptor pathway in prostate cancer. Curr Opin Pharmacol 8:440–448

Chiang C, Chiu M, Moore AJ, Anderson PH, Ghasem-Zadeh A, McManus JF, Ma C, Seeman E, Clemens TL, Morris HA, Zajac JD, Davey RA (2009) Mineralization and bone resorption are regulated by the androgen receptor in male mice. J Bone Miner Res 24:621–631

Choong CS, Wilson EM (1998) Trinucleotide repeats in the human androgen receptor: a molecular basis for disease. J Mol Endocrinol 21:235–257

Cloke B, Christian M (2012) The role of androgens and the androgen receptor in cycling endometrium. Mol Cell Endocrinol 358:166–175

Cloke B, Huhtinen K, Fusi L, Kajihara T, Yliheikkila M, Ho KK, Teklenburg G, Lavery S, Jones MC, Trew G, Kim JJ, Lam EW, Cartwright JE, Poutanen M, Brosens JJ (2008) The androgen and progesterone receptors regulate distinct gene networks and cellular functions in decidualizing endometrium. Endocrinology 149:4462–4474

Crocoll A, Zhu CC, Cato AC, Blum M (1998) Expression of androgen receptor mRNA during mouse embryogenesis. Mech Dev 72:175–178

Cunha GR, Donjacour A (1987) Stromal-epithelial interactions in normal and abnormal prostatic development. Prog Clin Biol Res 239:251–272

D'Antonio JM, VanderGriend DJ, Antony L, Ndikuyeze G, Dalrymple SL, Koochekpour S, Isaacs JT (2010) Loss of androgen receptor-dependent growth suppression by prostate cancer cells can occur independently from acquiring oncogenic addiction to androgen receptor signalling. PLoS ONE 5:e11475

De Gendt K, Verhoeven G (2012) Tissue- and cell-specific functions of the androgen receptor revealed through conditional knockout models in mice. Mol Cell Endocrinol 352(1–2):13–25

Deanesly R, Parkes AS (1933) Size changes in the seminal vesicles of the mouse during development and after castration. J Physiol 78:442–450

Deanesly R, Parkes AS (1936) Comparative activities of compounds of the androsterone-testosterone series. Biochem J 30:291–303

Dehm SM, Tindall DJ (2011) Alternatively spliced androgen receptor variants. Endocr Relat Cancer 18:83–96

Dubois V, Laurent M, Boonen S, Vanderschueren D, Claessens F (2012) Androgens and skeletal muscle: cellular and molecular action mechanisms underlying the anabolic actions. Cell Mol Life Sci 69:1651–1667

English KM, Steeds R, Jones TH, Channer KS (1997) Testosterone and coronary heart disease: is there a link? QJM 90:787–791

Fan X, Robaire B (1998) Orchidectomy induces a wave of apoptotic cell death in the epididymis. Endocrinology 139:2128–2136

Fawcett DW, Hoffer AP (1979) Failure of exogenous androgen to prevent regression of the initial segments of the rat epididymis after efferent duct ligation or orchidectomy. Biol Reprod 20:162–181

Ferlay J, Shin HR, Bray F, Forman D, Mathers C, Parkin DM (2010) Estimates of worldwide burden of cancer in 2008: GLOBOCAN 2008. Int J Cancer 127:2893–2917

George FW, Wilson J (1994) Gonads and ducts in mammals. In: Knobil E, Neill JD (eds) The physiology of reproduction, 2nd edn. Raven Press, New York, pp 3–27

Greene KL, Li LC, Okino ST, Carroll PR (2008) Molecular basis of prostate cancer. In: Memdelson J, Howley PM, Israel MA, Gray JA, Thompson CB (ed) The molecular basis of cancer, 3rd Edn

Haendler B, Cleve A (2012) Recent developments in antiandrogens and selective androgen receptor modulators. Mol Cell Endocrinol 352:79–91

Hanke H, Lenz C, Hess B, Spindler KD, Weidemann W (2001) Effect of testosterone on plaque development and androgen receptor expression in the arterial vessel wall. Circulation 103:1382–1385

Hay CW, McEwan IJ (2012) The impact of point mutations in the human androgen receptor: classification of mutations on the basis of transcriptional activity. PLoS ONE 7:e32514

He WW, Kumar MV, Tindall DJ (1991) A frame-shift mutation in the androgen receptor gene causes complete androgen insensitivity in the testicular-feminized mouse. Nucleic Acids Res 19:2373–2378

Hickey TE, Robinson JL, Carroll JS, Tilley WD (2012) Minireview: the androgen receptor in breast tissues: growth inhibitor, tumor suppressor, oncogene? Mol Endocrinol 26:1252–1267

Huggins C, Hodges CV (1941) Studies on prostatic cancer: the effect of castration, of estrogen and of androgen injection on serum phosphatases in metastatic carcinoma of the prostate. J Urol 16:9–12

Imai Y, Kondoh S, Kouzmenko A, Kato S (2009) Regulation of bone metabolism by nuclear receptors. Mol Cell Endocrinol 310(1–2):3–10

Ikeda Y, Aihara K, Yoshida S, Sato T, Yagi S et al (2009) Androgen-androgen receptor system protects against angiotensin II-induced vascular remodeling. Endocrinology 150:2857–2864

Ikeda Y, Aihara K, Akaike M, Sato T, Ishikawa K et al (2010) Androgen receptor counteracts Doxorubicin-induced cardiotoxicity in male mice. Mol Endocrinol 24:1338–1348

Jones RD, Pugh PJ, Hall J, Channer KS, Jones TH (2003) Altered circulating hormone levels, endothelial function and vascular reactivity in the testicular feminised mouse. Eur J Endocrinol 148:111–120

Jordan CL, Lieberman AP (2008) Spinal and bulbar muscular atrophy: a motoneuron or muscle disease? Curr Opin Pharmacol 8:752–758

Kaushik M, Sontineni SP, Hunter C (2010) Cardiovascular disease and androgens: a review. Int J Cardiol 142:8–14

Kawano H, Sato T, Yamada T, Matsumoto T, Sekine K, Watanabe T, Nakamura T, Fukuda T, Yoshimura K, Yoshizawa T, Aihara K, Yamamoto Y, Nakamichi Y, Metzger D, Chambon P, Nakamura K, Kawaguchi H, Kato S (2003) Suppressive function of androgen receptor in bone resorption. Proc Natl Acad Sci U S A 100:9416–9421

Khaw KT, Dowsett M, Folkerd E, Bingham S, Wareham N et al (2007) Endogenous testosterone and mortality due to all causes, cardiovascular disease, and cancer in men: European prospective investigation into cancer in Norfolk (EPIC-Norfolk) prospective population study. Circulation 116:2694–2701

Knudsen KE, Kelly WK (2011) Outsmarting androgen receptor: creative approaches for targeting aberrant androgen signaling in advanced prostate cancer. Expert Rev Endocrinol Metab 6:483–493

Macleod KF (2010) The RB tumor suppressor: a gatekeeper to hormone independence in prostate cancer? J Clin Invest 120(12):4179–4182

La Spada AR, Wilson EM, Lubahn DB, Harding AE, Fischbeck KH (1991) Androgen receptor gene mutations in X-linked spinal and bulbar muscular atrophy. Nature 352:77–79

Labrie F, Luu-The V, Labrie C, Belanger A, Simard J, Lin SX, Pelletier G (2003) Endocrine and intracrine sources of androgens in women: inhibition of breast cancer and other roles of androgens and their precursor dehydroepiandrosterone. Endocr Rev 24:152–182

Lamont KR, Tindall DJ (2011) Minireview: alternative activation pathways for the androgen receptor in prostate cancer. Mol Endocrinol 25:897–907

Liu PY, Death AK, Handelsman DJ (2003) Androgens and cardiovascular disease. Endocr Rev 24:313–340

Liu PY, Christian RC, Ruan M, Miller VM, Fitzpatrick LA (2005) Correlating androgen and estrogen steroid receptor expression with coronary calcification and atherosclerosis in men without known coronary artery disease. J Clin Endocrinol Metab 90:1041–1046

Krutskikh A, De Gendt K, Sharp V, Verhoeven G, Poutanen M, Huhtaniemi I (2010) Targeted inactivation of the androgen receptor gene in murine proximal epididymis causes epithelial hypotrophy and obstructive azoospermia. Endocrinology 152:689–696

MacLean HE, Handelsman DJ (2009) Unraveling androgen action in muscle: genetic tools probing cellular mechanisms. Endocrinology 150:3437–3439

Mahendroo MS, Cala KM, Hess DL, RussellDW (2001) Unexpected virilization in male mice lacking steroid 5α-reductase enzymes. Endocrinology 142:4652–4662

Marker PC, Donjacour AA, Dahiya R, Cunha GR (2003) Hormonal, cellular, and molecular control of prostatic development. Dev Biol 253:165–174

Marsh JD, Lehmann MH, Ritchie RH, Gwathmey JK, Green GE et al (1998) Androgen receptors mediate hypertrophy in cardiac myocytes. Circulation 98:256–261

Marshall FF, ReinerWG Goldberg BS (1979) The embryologic origin of the caput epididymidis in the rat. Invest Urol 17:78–82

Marshall E, Lowrey J, MacPherson S, Maybin JA, Collins F, Critchley HO, Saunders PT (2011) In silico analysis identifies a novel role for androgens in the regulation of human endometrial apoptosis. J Clin Endocrinol Metab 96:4E1746–55

McEwan IJ (2001) Structural and functional alterations in the androgen receptor in spinal bulbar muscular atrophy. Biochem Soc Trans 29:222–227

McEwan IJ, McGuinness D, Hay CW, Millar RP, Saunders PT, Fraser HM (2010) Identification of androgen receptor phosphorylation in the primate ovary in vivo. Reproduction 140:93–104

McPhaul MJ (2002) Androgen receptor mutations and androgen insensitivity. Mol Cell Endocrinol 198:61–67

Memarzadeh S, Cai H, Janzen DM, Xin L, Lukacs R, Riedinger M, Zong Y, DeGendt K, Verhoeven G, Huang J, Witte ON (2011) Role of autonomous androgen receptor signaling in prostate cancer initiation is dichotomous and depends on the oncogenic signal. Proc Natl Acad Sci U S A 108:7962–7967

Monks DA, Rao P, Mo K, Johansen JA, Lewis G, Kemp MQ (2008) Androgen receptor and Kennedy disease/spinal bulbar muscular atrophy. Horm Behav 53:729–740

Mooradian AD, Morley JE, Korenman SG (1987) Biological actions of androgens. Endocr Rev 8:1–28

Murphy L, O'Shaughnessy PJ (1991) Testicular steroidogenesis in the testicular feminized (Tfm) mouse: loss of 17 alpha-hydroxylase activity. J Endocrinol 131:443–449

Naghi JJ, Philip KJ, DiLibero D, Willix R, Schwarz ER (2011) Testosterone therapy: treatment of metabolic disturbances in heart failure. J Cardiovasc Pharmacol Ther 16:14–23

Narayanan R, Mohler ML, Bohl CE, Miller DD, Dalton JT (2008) Selective androgen receptor modulators in preclinical and clinical development. Nucl Recept Signal 6:e010

Nathan L, Shi W, Dinh H, Mukherjee TK, Wang X et al (2001) Testosterone inhibits early atherogenesis by conversion to estradiol: critical role of aromatase. Proc Natl Acad Sci U S A 98:3589–3593

Nettleship JE, Jones TH, Channer KS, Jones RD (2007) Physiological testosterone replacement therapy attenuates fatty streak formation and improves high-density lipoprotein cholesterol in the Tfm mouse: an effect that is independent of the classic androgen receptor. Circulation 116:2427–2434

Niu Y, Altuwaijri S, Lai KP, Wu CT, Ricke WA, Messing EM, Yao J, Yeh S, Chang C (2008a) Androgen receptor is a tumor suppressor and proliferator in prostate cancer. Proc Natl Acad Sci U S A 105:12182–12187

Niu Y, Altuwaijri S, Yeh S, Lai KP, Yu S, Chuang KH, Huang SP, Lardy H, Chang C (2008b) Targeting the stromal androgen receptor in primary prostate tumors at earlier stages. Proc Natl Acad Sci U S A 105:12188–12193

Notelovitz M (2002) Androgen effects on bone and muscle. Fertil Steril 77(Suppl 4):S34–S41

Notini AJ, McManus JF, Moore A, Bouxsein M, Jimenez M, Chiu WS, Glatt V, Kream BE, Handelsman DJ, Morris HA, Zajac JD, Davey RA (2007) Osteoblast deletion of exon 3 of the androgen receptor gene results in trabecular bone loss in adult male mice. J Bone Miner Res 22:347–356

O'Hara L, Welsh M, Saunders PTK, Smith LB (2010) Androgen receptor expression in the caput epididymal epithelium is essential for development of the initial segment and epididymal spermatozoa transit. Endocrinology 152:718–729

O'Hara L, Smith LB (2012) Androgen receptor signalling in Vascular Endothelial cells is dispensable for spermatogenesis and male fertility. BMC Res Notes 5:16

Oldridge EE, Pellacani D, Collins AT, Maitland NJ (2012) Prostate cancer stem cells: are they androgen-responsive? Mol Cell Endocrinol 360:14–24

Palazzolo I, Gliozzi A, Rusmini P, Sau D, Crippa V, Simonini F, Onesto E, Bolzoni E, Poletti A (2008) The role of the polyglutamine tract in androgen receptor. J Steroid Biochem Mol Biol 108:245–253

Pang SF, Chow PH, Wong TM (1979) The role of the seminal vesicles, coagulating glands and prostate glands on the fertility and fecundity of mice. J Reprod Fertil 56:129–132

Peitz B, Olds-Clarke P (1986) Effects of seminal vesicle removal on fertility and uterine sperm motility in the house mouse. Biol Reprod 35:608–617

Poletti A (2004) The polyglutamine tract of androgen receptor: from functions to dysfunctions in motor neurons. Front Neuroendocrinol 25:1–26

Practice Committee of American Society for Reproductive Medicine in collaboration with Society for Male Reproduction and Urology (2008) The management of infertility due to obstructive azoospermia. Fertil Steril 90:S121–S124

Presbitero P, Belli G, Zavalloni D, Rossi ML, Lisignoli V et al (2008) "Gender paradox" in outcome after percutaneous coronary intervention with paclitaxel eluting stents. EuroIntervention 4:345–350

Welsh M, Moffat L, McNeilly A, Brownstein D, Saunders PT, Sharpe RM, Smith LB (2011) Smooth muscle cell-specific knockout of androgen receptor: a new model for prostatic disease. Endocrinology 152:3541–3551

Regitz-Zagrosek V (2006) Therapeutic implications of the gender-specific aspects of cardiovascular disease. Nat Rev Drug Discov 5:425–438

Robaire B, Ewing LL, Zirkin BR, Irby DC (1977) Steroid delta4-5alpha-reductase and 3alpha-hydroxystreoid dehydrogenase in the rat epididymis. Endocrinology 101:1379–1390

Rosano GM, Leonardo F, Pagnotta P, Pelliccia F, Panina G et al (1999) Acute anti-ischemic effect of testosterone in men with coronary artery disease. Circulation 99:1666–1670

Rossouw JE, Anderson GL, Prentice RL, LaCroix AZ, Kooperberg C et al (2002) Risks and benefits of estrogen plus progestin in healthy postmenopausal women: principal results From the Women's Health Initiative randomized controlled trial. JAMA 288:321–333

Schulman C, Irani J, Aapro M (2012) Improving the management of patients with prostate cancer receiving long-term androgen deprivation therapy. BJU Int 109(Suppl 6):13–21

Seth A, Serruys PW, Lansky A, Hermiller J, Onuma Y et al (2010) A pooled gender based analysis comparing the XIENCE V(R) everolimus-eluting stent and the TAXUS paclitaxel-eluting stent in male and female patients with coronary artery disease, results of the SPIRIT II and SPIRIT III studies: two-year analysis. EuroIntervention 5:788–794

Shen MM, Abate-Shen C (2010) Molecular genetics of prostate cancer: new prospects for old challenges. Genes Dev 24:1967–2000

Simanainen U, McNamara K, Davey RA, Zajac JD, Handelsman DJ (2008) Severe subfertility in mice with androgen receptor inactivation in sex accessory organs but not in testis. Endocrinology 149:3330–3338

Sinha-Hikim I, Taylor WE, Gonzalez-Cadavid NF, Zheng W, Bhasin S (2004) Androgen receptor in human skeletal muscle and cultured muscle satellite cells: up-regulation by androgen treatment. J Clin Endocrinol Metab 89:5245–5255

Sinnesael M, Boonen S, Claessens F, Gielen E, Vanderschueren D (2011) Testosterone and the male skeleton: a dual mode of action. J Osteoporos 2011:240328

Son BK, Akishita M, Iijima K, Ogawa S, Maemura K et al (2010) Androgen receptor-dependent transactivation of growth arrest-specific gene 6 mediates inhibitory effects of testosterone on vascular calcification. J Biol Chem 285:7537–7544

Sullivan ML, Martinez CM, Gennis P, Gallagher EJ (1998) The cardiac toxicity of anabolic steroids. Prog Cardiovasc Dis 41:1–15

Takagi-Morishita Y, Kuhara A, Sugihara A, Yamada N, Yamamoto R, Iwasaki T, Tsujimura T, Tanji N, Terada N (2002) Castration induces apoptosis in the mouse epididymis during postnatal development. Endocr J 49:75–84

Tanaka F, Katsuno M, Banno H, Suzuki K, Adachi H, Sobue G (2012) Current status of treatment of spinal and bulbar muscular atrophy. Neural Plast 2012:369284

Tatchum-Talom R, Martel C, Marette A (2002) Effects of ethinyl estradiol, estradiol, and testosterone on hindlimb endothelial function in vivo. J Cardiovasc Pharmacol 39:496–502

Teoh H, Quan A, Man RY (2000) Acute impairment of relaxation by low levels of testosterone in porcine coronary arteries. Cardiovasc Res 45:1010–1018

Tharp DL, Masseau I, Ivey J, Ganjam VK, Bowles DK (2009) Endogenous testosterone attenuates neointima formation after moderate coronary balloon injury in male swine. Cardiovasc Res 82:152–160

Tran C, Ouk S, Clegg NJ, Chen Y, Watson PA, Arora V, Wongvipat J, Smith-Jones PM, Yoo D, Kwon A, Wasielewska T, Welsbie D, Chen CD, Higano CS, Beer TM, Hung DT, Scher HI, Jung ME, Sawyers CL (2009) Development of a second-generation antiandrogen for treatment of advanced prostate cancer. Science 324:787–790

Travison TG, Araujo AB, Kupelian V, O'Donnell AB, McKinlay JB (2007) The relative contributions of aging, health, and lifestyle factors to serum testosterone decline in men. J Clin Endocrinol Metab 92:549–555

Travison TG, Araujo AB, Hall SA, McKinlay JB (2009) Temporal trends in testosterone levels and treatment in older men. Curr Opin Endocrinol Diabetes Obes 16:211–217

Waltering KK, Urbanucci A, Visakorpi T (2012) Androgen receptor (AR) aberrations in castration-resistant prostate cancer. Mol Cell Endocrinol 360:38–43

Walters KA, Allan CM, Handelsman DJ (2008) Androgen actions and the ovary. Biol Reprod 78:380–389

Wang Y, Romigh T, He X, Tan MH, Orloff MS, Silverman RH, Heston WD, Eng C (2011) Differential regulation of PTEN expression by androgen receptor in prostate and breast cancers. Oncogene 30:4327–4338

Welsh M, Saunders PT, Marchetti NI, Sharpe RM (2006) Androgen dependent mechanisms of Wolffian duct development and their perturbation by flutamide. Endocrinology 147:4820–4830

Welsh M, Sharpe RM, Moffat L, Atanassova N, Saunders PT et al (2010) Androgen action via testicular arteriole smooth muscle cells is important for Leydig cell function, vasomotion and testicular fluid dynamics. PLoS ONE 5:e13632

Wilson JD, George FW, Griffin JE (1981) The hormonal control of sexual development. Science 211:1278–1284

Wilson EM (2007) Muscle-bound? A tissue-selective nonsteroidal androgen receptor modulator. Endocrinol 148(1):1–3

Wingard DL, Suarez L, Barrett-Connor E (1983) The sex differential in mortality from all causes and ischemic heart disease. Am J Epidemiol 117:165–172

Wu CT, Altuwaijri S, Ricke WA, Huang SP, Yeh S, Zhang C, Niu Y, Tsai MY, Chang C (2007) Increased prostate cell proliferation and loss of cell differentiation in mice lacking prostate epithelial androgen receptor. Proc Natl Acad Sci U S A 104:12679–12684

Wu FC, Tajar A, Pye SR, Silman AJ, Finn JD et al (2008) Hypothalamic-pituitary-testicular axis disruptions in older men are differentially linked to age and modifiable risk factors: the European male aging study. J Clin Endocrinol Metab 93:2737–2745

Wyce A, Bai Y, Nagpal S, Thompson CC (2010) Research resource: the androgen receptor modulates expression of genes with critical roles in muscle development and function. Mol Endocrinol 24:1665–1674

Yeh S et al (2002) Generation and characterization of androgen receptor knockout (ARKO) mice: an in vivo model for the study of androgen functions in selective tissues. Proc Natl Acad Sci U S A 99(21):13498–13503

Yu J, Akishita M, Eto M, Ogawa S, Son BK et al (2010) Androgen receptor-dependent activation of endothelial nitric oxide synthase in vascular endothelial cells: role of phosphatidylinositol 3-kinase/akt pathway. Endocrinology 151:1822–1828

Chapter 6
Manipulating Androgens for Therapy

Androgens are important during all stages of development from fetal life to adulthood. There are many clinical situations in which manipulation of androgens may be required. This generally involves either increasing androgen levels in patients who are deficient or suppressing androgens in patients with conditions of androgen excess.

Androgen action is important in fetal life to masculinize the male fetus. Failure of masculinization can occur in many Disorders of Sex Development (DSD). Some of these disorders result in ambiguous genitalia at birth due to virilization of a female fetus or under-virilization of a male fetus (see Chap. 2). Androgens are also produced at relatively high levels in the immediate postnatal period during the so-called 'mini-puberty'. The importance of this period of androgen production in the male is not completely understood. Androgen levels then remain low during the childhood period until the onset of puberty when production of androgens is important for the development of secondary sexual characteristics and spermatogenesis.

6.1 Management of DSD

There are many DSD conditions that require manipulation of androgens. These can be classified into disorders of androgen production, lack of androgen action or failure of peripheral conversion of androgens (Table 6.1). Here, we describe conditions that result in each of these situations.

6.1.1 Congenital Adrenal Hyperplasia

Congenital adrenal hyperplasia results from mutations in enzymes involved in the steroidogenic pathway. The most common mutation is in the gene for 21-hydroxylase (95 % of CAH) which will be discussed here although there are other mutations that may result in variations on the 21-hydroxylase phenotype. The

L. B. Smith et al., *Testosterone: From Basic Research to Clinical Applications*, SpringerBriefs in Reproductive Biology, DOI: 10.1007/978-1-4614-8978-8_6, © The Author(s) 2013

Table 6.1 Disorders of sex development (DSD) that arise from abnormalities in androgen synthesis, action, or conversion

	Disorder	Karyotype	Clinical phenotype
Androgen excess	Placental aromatase deficiency	XX	Virilized female
- Maternal	Maternal drugs (androgens)	XX	Virilized female
	Androgen secreting tumors	XX	Virilized female
Androgen excess	CAH (e.g., CYP21A2, HSD3B2)	XX	Virilized female
- Fetal	Glucocorticoid receptor mutation	XX	Virilized female
Androgen deficiency	LH receptor mutation	XY	Undervirilized male
	CAH (e.g., HSD3B2)	XY	Undervirilized male
	Smith Lemli Opitz syndrome	XY	Undervirilized male
Androgen action	Androgen insensitivity syndrome		
	- Complete (CAIS)	XY	Female
	- Partial (PAIS)	XY	Undervirilized male
Peripheral conversion	5α-reductase deficiency	XY	Undervirilized male

The karyotype and associated clinical phenotype associated with these disorders are also described. CAH—Congenital adrenal hyperplasia

result of this mutation is a block in the production of glucocorticoids and mineralocorticoid. This in turn results in a lack of negative feedback on the hypothalamo-pituitary axis by glucocorticoids and therefore excess ACTH is produced in response. The high levels of ACTH stimulate the production of excess androgens from the adrenal gland resulting in virilization in patients.

CAH may present in the neonatal period in females as ambiguous genitalia, or in males with a 'salt wasting crisis', resulting from deficiency in mineralocorticoid ('classic salt-wasting CAH'). In childhood it may present as virilization in males and females with appearance of pubic hair, cliteromegaly or penile enlargement and tall stature ('classical simple virilising'). Rarely it may present in adulthood with hirsutism, infertility, or oligomenorrhea in women ('non classical') (Huynh et al. 2009).

The goal of treatment in patients with CAH is to reduce the excessive androgen production, while at the same time replacing the glucocorticoid and mineralocorticoid deficiency. Both of these aims are achieved by treatment with glucocorticoid, most frequently in the form of oral hydrocortisone tablets three times a day, with dosages aimed to mimic the physiological diurnal rhythm of cortisol production. Higher dose are generally required in the morning surge in glucocorticoid production, while lower doses are required in the afternoon as there is a physiological dip in glucocorticoid during this period of the day. Mineralocorticoids are also replaced usually with twice daily oral medication. Patients are at risk of acute adrenal insufficiency if they fail to take their medication or they become unwell. This is a medical emergency and requires immediate treatment with high dose intravenous hydrocortisone and rehydration. Patients are advised to double or triple their oral steroid doses when they are unwell to avoid this complication.

In children, optimal treatment is required for maintaining normal growth and development. The dose of steroid must be carefully tailored to avoid the problems associated with deficient or excess glucocorticoid. Overtreatment will result in glucocorticoid excess and the complications associated with Cushing syndrome. This includes growth failure and short stature. The opposite occurs with under-treatment of CAH, where the excess androgens stimulate growth and result in tall stature. This may also predispose to precocious puberty which will ultimately result in short stature due to premature epiphyseal closure. In patients who have gone into precocious puberty, treatment to suppress puberty with GnRH agonists may be required. Careful assessment of growth is required using standard growth charts appropriate for the population in addition to frequent assessment of bone age using an X-ray of the left wrist and assessment of puberty using Tanner staging.

CAH may be suspected during fetal life in mothers who have had an affected child previously. Treatment to prevent virilization of females in utero involves administering dexamethasone to the mother in order to restore the negative feedback on the hypothalamo-pituitary axis and switch off the ACTH drive of androgen production. This therapy although effective in reducing virilization in affected females is controversial for a number of reasons including the fact that for every affected female there are seven fetuses that are exposed to dexamethasone in utero, including the unaffected females and all of the males. This treatment may be associated with long-term cognitive effects in both males and females (Kim 2012). Improvements in karyotyping during pregnancy using fetal cells/DNA from maternal blood may help to reduce the length of treatment in boys (Simpson 2011). It is therefore currently recommended that in utero treatment for CAH should be performed only as part of clinical research studies.

6.1.2 Androgen Insensitivity Syndrome

Androgen Insensitivity syndrome (AIS) is a condition resulting from mutation in the AR. The mutation may result in Complete AIS (CAIS) or alternatively if there is some residual function of the androgen receptor then this is termed Partial AIS (PAIS). Testosterone is produced by the testis and can also be converted to DHT, but the signaling in the peripheral tissues is reduced due to the reduced/absent AR function. The result is failure of normal masculinization in males. Individuals with PAIS may present in infancy with a variable degree of ambiguous genitalia in an XY baby. Depending on the degree of phenotypic appearance and the biochemical profile these patients may be raised as male or female. The process involved in assigning gender is beyond the scope of this chapter but will undoubtedly require the involvement of a multidisciplinary team including an endocrinologist, urologist, and psychologist (Hughes 2012). If, as the majority are, the patient is raised male, then treatment may involve testosterone therapy from adolescence in order to try and overcome the reduced function of the AR. DHT therapy may also be considered as this will not be aromatized to estradiol and will not contribute further to the gynecomastia that occurs in these patients. Treatment with androgens for pubertal induction is not always required. No clear consensus on the management of these patients exists which may be due to the heterogeneity of the clinical phenotype and therefore treatment must be individualized. There is also an increased risk of testicular cancer and these patients require at the very least orchidopexy with frequent clinical monitoring, or alternatively the gonads may be surgically removed. Surgery will also be required for any associated hypospadias in childhood and many patients with gynecomastia also require breast reduction surgery in adolescence.

Patients with CAIS present as phenotypically female. This condition may be identified during surgery for inguinal hernia which turns out to contain a testis. Alternatively it can be diagnosed during investigation for amenorrhea. The testis produces AMH which induces regression of the Müllerian structures and affected patients therefore lack a uterus, fallopian tubes, and upper two-thirds of the vagina. Testosterone levels are often increased and aromatization of testosterone to estradiol results in breast development at puberty. Treatment of these patients involves hormone replacement in the form of exogenous estrogen either for pubertal induction or for adult maintenance therapy (Hughes 2012). The gonads should also be removed as they have an increased risk of malignancy, however, the timing of removal should be considered. The tumor risk is low before puberty and therefore in some cases the gonads may be left until adolescence in order to permit the progression of puberty. If the gonads are removed in infancy/childhood then pubertal induction with increasing doses of estrogen will be required. Once puberty is complete, patients can then receive an adult dose of estrogen. There are many preparations including oral (e.g., oral contraceptive pill; OCP) and transdermal, and given that these patients have no uterus they do not require progesterone replacement. Infertility is an inevitable consequence of this disorder and psychology support regarding the diagnosis and the resulting infertility is an important part of the management in these patients.

6.1.3 5α-Reductase Deficiency

5α-reductase Deficiency is a rare condition that is caused by a mutation in the enzyme that converts testosterone to the more potent androgen DHT. This results in ambiguous genitalia in XY patients at birth. The phenotype may range from female external genitalia to male appearance with hypospadias and/or micropenis. This is due to the fact that DHT is important for masculinization of the external genitalia. Despite these features at birth, patients raised male usually undergo a normal puberty, most likely due to the fact that conversion of testosterone to DHT at this stage involves a different isoenzyme of 5α-reductase and also because testosterone is the predominant determinant of growth of the external genitalia and development of male secondary sexual characteristics during puberty (Cheon 2011).

An important consideration in these patients is gender assignment. Previously the majority of patients were raised female as the condition was usually picked up in those who were phenotypically closer to female than male. However, there was a high rate of gender identity change to male in these patients which may be related to masculinization of the brain by testosterone (Cohen-Kettenis 2005). Diagnosis may be made on the basis of a raised T:DHT (>30:1) ratio following hCG stimulation during infancy or at puberty. Treatment consists of surgical correction of crypt-orchidism to optimize future spermatogenesis, along with hypospadias repair, which may be preceded by topical application of DHT cream to improve phallic length prior to surgery (Carrillo 2007). There remains a tumor risk in these patients although it is believed to be low risk, nevertheless the testes should be monitored at regular intervals. If the patient is to be raised female then corrective surgery may be performed to reconstruct the external genitalia and create a vaginal opening. In this situation, gonads should be removed to prevent virilization at puberty and the risk of malignancy (Cheon 2011).

6.2 Delayed Puberty and Pubertal Induction

Delayed puberty in males is defined as failure to enter puberty by the age of 14 years. There are many conditions in the male that result in delayed puberty. Most commonly this is due to constitutional delay of growth and puberty (CDGP). In these patients there is no underlying pathology. Frequently there is a family history of pubertal delay in first-degree relatives. These patients require assessment of pubertal staging using the Tanner scale and accurate measurement of growth using appropriate growth charts. Bone age X-ray should also be performed, which will reveal a delay in patients with CDGP and therefore reassure most patients that they are likely to achieve their target height. Puberty will eventually progress but these patients may require a short course of treatment with testosterone either by monthly long-acting intramuscular injection, topical gel or alternatively with daily oral testosterone, to induce some secondary sexual characteristics and short-term growth benefits. Although there are numerous preparations of testosterone available, these are the only options that can be given in appropriate dosage to adolescents for puberty induction (Rogol 2005). An increase in height velocity can also be achieved in boys with CDGP using the anabolic steroid oxandrolone, however, this does not result in secondary sexual characteristics that many boys wish to attain. In addition, availability of this medication may also be limited. Testosterone therapy may be given for 3–6 months in the first instance (further doses may be given). Although it is not clear whether these short courses of testosterone alter the timing of puberty it may offer some psychological benefit to patients if they are concerned about their lack of pubertal development. Testicular volume should be assessed at regular intervals to ensure that puberty is initiated. Testicular enlargement (>3mls) indicates that the patient is in puberty. Treatment can be stopped once the patient is established in puberty as further courses of testosterone are not likely to provide any further benefit over endogenous testosterone when the testicular volume has reached 10 ml. The goal of therapy is to achieve linear growth and development of secondary sexual characteristics, while allowing the accrual of muscle mass and bone mineral content (Rogol 2005). If the testes fail to enlarge then a diagnosis of hypogonadotrophic hypogonadism may be considered. These patients will benefit from continued testosterone therapy in gradually increasing dosage to complete puberty (~ 2 years) followed by an adult maintenance dose of testosterone for which there are numerous preparations available, including transdermal patches, gels, and very long acting injectable testosterone. In patients with hypogonadotrophic hypogonadism, hCG $\pm$ FSH may also be an option for induction of puberty, however, from a practical point of view it can be extremely difficult to distinguish IHH from CDGP and therefore testosterone is the most commonly used method of pubertal induction.

Patients with primary gonadal failure will also require testosterone replacement. This group of patients most commonly consists of patients with Klinefelter Syndrome or those who have received gonadotoxic therapies (e.g., cancer therapy).

6.3 Fertility Induction

Testosterone therapy is extremely effective in inducing puberty changes in adolescence and also in maintaining male sexual function; however, exogenous testosterone therapy does not result in spermatogenesis and fertility. Patients requiring fertility need to be able to generate intra-testicular testosterone. Spermatogenesis may be induced in patients with secondary gonadal failure by treatment with either GnRH (pulsatile) or gonadotrophins, however, patients with hypopituitarism will not respond to GnRH therapy (Büchter 1998). Treatment with GnRH or a combination of hCG and human menopausal gonadotrophins (hMG; consisting of FSH, LH, and hCG) has been shown to result in a doubling of testicular volume over 3–5 months and the presence of sperm in the majority of patients, leading to a high frequency of pregnancies (Buchter 1998). More recently, recombinant preparations of FSH have become available and can be used in combination to hCG for induction of spermatogenesis in hypogonadal men (Warne 2009).

6.4 Erectile Dysfunction

Erectile dysfunction (ED) is defined as the failure to attain or maintain a penile erection sufficient for successful vaginal intercourse (Anonymous 1993). There are a number of conditions associated with ED, the most common of which is type 2 diabetes mellitus, however, the major risk factor is age. Other predictors include lifestyle factors such as obesity, excessive alcohol use, and lack of physical

exercise. The causes may be classified into psychogenic, neurogenic, endocrino-
logical, or vasculogenic (Shoumloul 2013) and the category may determine the
treatments that are most likely to be effective. Modification of lifestyle and use of
phosphodiesterase Type 5 (PDE-5) inhibitors (e.g., sildenafil) may be considered
first-line therapy (Hatzimouratidis 2010). Reduction in weight with increase in
physical activity has been proven to be effective, while the use of PDE-5 inhibitors
have revolutionized the treatment of ED in the past 20 years with the use of these
medications in the primary care setting (Tsertsvadze 2009). Androgens are known
to play an important role in erectile function including enhancing sexual desire and
regulation of nitric oxide synthase and PDE-5 expression, both important factors in
erectile function (Shoumoul 2013). Initial investigation should include measurement of serum testosterone and if low should be followed up by measuring
gonadotrophins and prolactin, both of which can influence testosterone production.
Despite the importance of testosterone in erectile function, the use of exogenous
testosterone is currently only recommended in patients with primary gonadal
failure. A meta-analysis has demonstrated beneficial effects on erectile function in
these patients (Jain 2000), while a more recent study demonstrated significant
effects of testosterone replacement on satisfaction with erectile function in patients
with a low testosterone (Boloña 2007). The Clinical Practice Guidelines of the
Endocrine Society also state that testosterone should be offered to patients with
ED, only when the serum testosterone levels are unequivocally low (Bhasin 2010).

6.5   Testosterone in the Aging Male: The 'Andropause'

Aging is associated with a decline in sex hormones in both men and women. In
men the decline in testosterone is termed the 'andropause' and this is associated
with a decline in bone/muscle mass and physical function (Horstman 2012). The
reduced muscle mass is associated with a wide range of chronic conditions such as
obesity, chronic obstructive pulmonary disease, type 2 diabetes mellitus, and also
increased all-cause mortality (Horstman 2012). A study of more than 1,000 men
demonstrated that from the age of 35–40 testosterone levels decrease by around
2 % per year (Feldman 2002). Treatment of 'testosterone deficiency' in older men
is still a contentious issue as there is a lack of good evidence for benefit and there
is an increased risk of adverse outcome (Bhasin 2010).

A Clinical Practice Guideline has been developed by the Endocrine Society and
this is summarized below (Bhasin 2010). The Guideline states that diagnosis of
testosterone deficiency is based on a morning plasma total testosterone level with a
repeat sample if the initial value is low because 30–35 % of men with a low
testosterone on initial sampling have a normal testosterone production over a 24 h
period (Swerdlow 2000). There must also be evidence of symptoms suggestive of
testosterone deficiency, except in conditions that are associated with low testosterone such as type II diabetes mellitus, osteoporosis, infertility, and chronic
obstructive pulmonary disease, where patients may be likely to benefit from

testosterone therapy (Bhasin 2010). Exclusion of pathological cause for a low testosterone should be undertaken with measurement of LH and FSH to establish whether this is primary or secondary hypogonadism, with further investigation as necessary (Bhasin 2010).

Treatment should be reserved for symptomatic patients with low testosterone levels. Care must be taken in older men due to the fact that testosterone therapy will have a negative impact in patients with prostate disease including metastatic prostate cancer, in which proliferation may be stimulated by testosterone (Fowler 1981) (see Chap. 5). In the case of prostate disease, e.g., nodules, induration, raised prostate-specific antigen (PSA), severe hyperplasia and cancer, testosterone therapy should not be initiated. Patients receiving high dose glucocorticoid therapy, who exhibit low testosterone levels may also receive testosterone replacement to protect bone health and maintain bone mineral density (Bhasin 2010). The goal of therapy is to give physiological replacement of testosterone and maintain the testosterone level in the mid-normal range. A wide variety of preparations are available and patients should be able to choose the option that best suits them. This may include gels, patches, injections, or oral formulations and administration frequency ranges from 12 h to 2 weeks depending on the preparation. Patients should be monitored regularly, e.g., at 3 months and annually thereafter (Bhasin 2010). Testosterone levels should be monitored and the timing of sampling in relation to the last testosterone dose is determined by the type of preparation used. Specific side-effects such as local irritation, alterations in taste and fluctuating libido are also dependent on the type of preparation used. Care must be taken to ensure that side effects such as polycythemia are detected and therapy stopped if the hematocrit is greater than 54 %. In addition, the patients should have their prostate examined and PSA levels measured and if the PSA velocity is increased this should prompt referral to an urologist (Carter 1997). Much of the evidence that is used to make the guideline for testosterone replacement in aging males is graded as 'very low' or 'low' quality and it is clear that a significant amount of research into this subject is required.

6.6 Polycystic Ovarian Syndrome (PCOS)

PCOS is the commonest condition in women of reproductive age, affecting 6–10 % of women, and classically consisting of the triad of menstrual disorders, polycystic ovaries, and hyper-androgenism (Broekmans 2006). PCOS can be considered a heterogeneous condition and as a result there are numerous definitions, however, the majority of definitions require at least two of the three parts of the triad (Broekmans 2006). PCOS is frequently associated with obesity and insulin resistance and leads to an increased risk of type II diabetes mellitus. The insulin resistance and hyperinsulinism results in effects on LH and IGF-1, which in turn lead to excessive ovarian androgen production (Goodarzi et al. 2011). This results in a vicious cycle, whereby excessive androgen production leads to reduced feedback inhibition of the hypothalamus and a resultant hypersecretion of LH and further increase in androgens (Goodarzi et al. 2011). In 50–75 % of women with PCOS there is an increase in circulating free and total testosterone or DHEAS, with increased free testosterone identified in 60 % of patients (Huang 2010).

In patients with suspected PCOS, diagnosis depends on the criteria mentioned above and the exclusion of other conditions including congenital adrenal hyperplasia, hyperprolactinemia, thyroid dysfunction, and rare androgen secreting tumors (Goodarzi et al. 2011). The aim of treatment is to control the symptoms and must be individualized for each patient. Central to the management of PCOS is lifestyle management and weight control. Weight gain worsens the reproductive and metabolic effects of PCOS and so weight loss in overweight and obese patients is an important aspect of treatment resulting in reduction of androgens and insulin and improving menstrual and ovulatory dysfunction (Goodarzi et al. 2011). For those with insulin resistance exercise and weight loss may be beneficial and an insulin sensitizer (e.g., Metformin) may also be used.

A common presenting complaint is hirsutism as a result of elevated androgens. The treatment involves a combination of pharmacological agents to reduce androgen secretion or action, and topical treatments to remove terminal hair (Escobar-Morreale 2012). Cosmetic hair removal may be helpful in mild cases, although effects may only be temporary. Electrolysis may give permanent results by destroying dermal papillae and laser therapy may also be used. Often cosmetic agents may prove useful in the short term as the pharmacological agents take several months before an improvement can be demonstrated (Escobar-Morreale 2012). Topical application of eflornithine cream can be used as a single treatment for mild cases of facial hirsutism or may be combined with other drug therapies. Oral medications can be considered in patients that are not wishing immediate fertility. Treatment should almost always include an agent for peripheral androgen blockade, combined with an oral contraceptive, as peripheral androgen blockers are teratogenic (Goodzari 2011). The OCP reduces ovarian androgen production via reduction in gonadotrophins, while increasing SHBG levels and thereby reducing free testosterone. The preparation of OCP may include an anti-androgen (e.g., cyproterone acetate) which may increase the anti-androgenic effect (Escobar-Morreale 2012). The effects of the various OCP preparations on thromboembolic risk and metabolic profile must also be taken into account. The OCP also has potential benefits in terms of regulating menstrual irregularity. Second line drug treatment involves the use of anti-androgens in combination with the OCP. The anti-androgen is either in the form of an AR blocker (flutamide, spironolactone, cyproterone acetate) or a 5α-reductase inhibitor (finasteride). Anti-androgens in combination with the OCP are more effective in improving hirsutism compared with the OCP alone, however, flutamide has been associated with a risk of severe liver toxicity and is not licensed for use in many countries.

For patients who have anovulatory cycles who require fertility, options include clomiphene citrate, metformin, or letrozole, all of which affect estrogen negative feedback to the hypothalamo-pituitary unit, resulting in an increase in gonadotrophin stimulation and induction of ovulation (Goodzari 2011).

6.7 Key Points

- Abnormalities in androgen production and action occur in a variety of conditions in both males and females.
- Indications for androgen therapy include primary and secondary gonadal failure in males.
- Androgen blockade may be required in females with conditions that result in excess androgen production (e.g., PCOS).

References

Anonymous (1993) NIH Consensus conference. impotence. nih consensus development panel on impotence. JAMA 270(1):83–90

Bhasin S, Cunningham GR, Hayes FJ, Matsumoto AM, Snyder PJ, Swerdloff RS, Montori VM (2010) Task force, endocrine society.testosterone therapy in men with androgen deficiency syndromes: an endocrine society clinical practice guideline. J Clin Endocrinol Metab 95(6):2536–2559

Bolo–a ER, Uraga MV, Haddad RM, Tracz MJ, Sideras K, Kennedy CC, Caples SM, Erwin PJ, Montori VM (2007) Testosterone use in men with sexual dysfunction: a systematic review and meta-analysis of randomized placebo-controlled trials. Mayo Clin Proc 82(1):20–8

Broekmans FJ, Knauff EA, Valkenburg O, Laven JS, Eijkemans MJ, Fauser BC (2006) PCOS according to the Rotterdam consensus criteria: change in prevalence among WHO-II anovulation and association with metabolic factors. BJOG 113(10):1210–1217

BŸchter D, Behre HM, Kliesch S, Nieschlag E (1998) Pulsatile GnRH or human chorionic gonadotropin/human menopausal gonadotropin as effective treatment for men with hypogonadotropic hypogonadism: a review of 42 cases. Eur J Endocrinol 3:298–303

Carter HB (1997) PSA variability versus velocity. Urology 49:305

Carrillo AA, Damian M, Berkovitz G (2007) Disorders of sexual differentiation. In: Lifshitz F (ed) Pediatric endocrinology, 5th edn. Marcel Dekker, New York, pp 365–390

Cheon CK (2011) Practical approach to steroid 5 alpha-reductase type 2 deficiency. Eur J Pediatr 170:1–8

Cohen-Kettenis PT (2005) Gender change in 46, XY persons with 5 alpha-reductase-2 deficiency and 17 beta-hydroxysteroid dehydrogenase-3 deficiency. Arch Sex Behav 34:399410

Escobar-Morreale HF, Carmina E, Dewailly D, Gambineri A, Kelestimur F, Moghetti P, Pugeat M, Qiao J, Wijeyaratne CN, Witchel SF, Norman RJ (2012) Epidemiology, diagnosis and management of hirsutism: a consensus statement by the Androgen Excess and Polycystic Ovary Syndrome Society. Hum Reprod Update 18(2):146–170

Feldman HA, Longcope C, Derby CA, Johannes CB, Araujo AB, Coviello AD, Bremner WJ, McKinlay JB (2002) Age trends in the level of serum testosterone and other hormones in middle-aged men: longitudinal results from the Massachusetts male aging study. J Clin Endocrinol Metab 87(2):589–598

Fowler JE Jr, Whitmore WF Jr (1981) The response of metastatic adenocarcinoma of the prostate to exogenous testosterone. J Urol 126:372–375

Goodarzi MO, Dumesic DA, Chazenbalk G, Azziz R (2011) Polycystic ovary syndrome: etiology, pathogenesis and diagnosis. Nat Rev Endocrinol 7(4):219–231

Hatzimouratidis K, Amar E, Eardley I et al European Association of Urology (2010) Guidelines on male sexual dysfunction: erectile dysfunction and premature ejaculation. Eur Urol 57:804–14

Horstman AM, Dillon EL, Urban RJ, Sheffield-Moore M (2012) The role of androgens and estrogens on healthy aging and longevity. J Gerontol A Biol Sci Med Sci 67(11):1140–1152

Huang A, Brennan K, Azziz R (2010) Prevalence of hyperandrogenemia in the polycystic ovary syndrome diagnosed by the National Institutes of Health 1990 criteria. Fertil Steril 93:1938–1941

Hughes IA, Davies JD, Bunch TI, Pasterski V, Mastroyannopoulou K, MacDougall J (2012) Androgen insensitivity syndrome. Lancet 380(9851):1419–1428

Huynh T, McGown I, Cowley D, Nyunt O, Leong GM, Harris M, Cotterill AM (2009) The clinical and biochemical spectrum of congenital adrenal hyperplasia secondary to 21-hydroxylase deficiency. Clin Biochem Rev 30(2):75–86

Jain P, Rademaker AW, McVary KT (2000) Testosterone supplementation for erectile dysfunction: results of a meta-analysis. J Urol 164:371–375

Kim MS, Ryabets-Lienhard A, Geffner ME (2012) Management of congenital adrenal hyperplasia in childhood. Curr Opin Endocrinol Diabetes Obes 19(6):483–488

Rogol AD (2005) New facets of androgen replacement therapy during childhood and adolescence. Expert Opin Pharmacother 6(8):1319–1336

Shoumoul R, Ganem H (2013) Erectile dysfunction. Lancet 381:153–65

Simpson JL, Bischoff F (2011) Novel non-invasive prenatal diagnosis as related to congenital adrenal hyperplasia. Adv Exp Med Biol 707:37–38

Swerdloff RS, Wang C, Cunningham G, Dobs A, Iranmanesh A, Matsumoto AM, Snyder PJ, Weber T, Longstreth J, Berman N (2000) Long-term pharmacokinetics of transdermal testosterone gel in hypogonadal men. J Clin Endocrinol Metab 85:4500–4510

Tsertsvadze A, Yazdi F, Fink HA et al (2009) Diagnosis and treatment of erectile dysfunction. Rockville (MD): agency for healthcare research and quality (US); 2009 May. (Evidence Reports/Technology Assessments, No 171) https://www.ncbi.nlm.nih.gov/books/NBK38725/

Warne DW, Decosterd G, Okada H, Yano Y, Koide N, Howles CM (2009) A combined analysis of data to identify predictive factors for spermatogenesis in men with hypogonadotropic hypogonadism treated with recombinant human follicle-stimulating hormone and human chorionic gonadotropin. Fertil Steril 92(2):594-604

Epilogue
Future Perspectives

From the very first investigations of what determines maleness carried out over 100 years ago, our understanding of the key functional roles played by testosterone has increased exponentially, and yet there is still much we do not understand.

The role of testosterone during the male programming window is now widely accepted, yet what controls the opening and closing of this window, so vital for male development remains unknown. We could speculate that it is the expression of AR that controls masculinization, and whilst this is a key factor, it cannot be the entire story, because presence of testosterone and AR at high levels outside of the MPW is unable to compensate for absence within the window. If expression of AR is a key player in this, we remain some way from fully understanding how this works as, surprisingly, what controls expression of AR is also still unknown. Information is already available about potential regulatory elements, but what will be required is a more comprehensive analysis of the AR gene in different cells and tissues to fully appreciate how this molecule is regulated. Establishing how AR expression and the MPW is controlled is a key objective of ongoing studies; given the devastating clinical impacts arising when this goes wrong, it is an important priority area for research.

How Androgens act also remains a key area for future research. Not only can androgens act directly via AR, but also through rapid non-genomic pathways, and through conversion to estradiol and activation of genomic (via estrogen receptors) and non-genomic estrogen-dependent pathways. Thus there are a myriad of signaling and transcriptional cascades impacted by changes in testosterone production and action. Dissecting this system apart requires further detailed investigation, possibly combining temporal as well as spatial manipulation of the androgen/AR signaling system, utilizing a combination of pharmaceutical and genetic manipulation.

Since the cloning of the androgen receptor cDNAs 25 years ago, considerable progress has been made in understanding the molecular mechanisms of AR action. One of the major challenges concern the conformation of the full-length receptor; it is clear the receptor can act as an allosteric switch for different bound ligands, including steroid, DNA or co-regulatory proteins, but a more complete understanding of how these are coordinated to give gene and cell-specific

L. B. Smith et al., *Testosterone: From Basic Research to Clinical Applications*,
SpringerBriefs in Reproductive Biology
DOI: 10.1007/978-1-4614-8978-8, © The Author(s) 2013

responses is just emerging. Similarly, it is known the receptor is a substrate for a plethora of post-translational modifications, including phosphorylation, acetylation, methylation, and sumolylation, but the functional consequences for specific modifications and how these are integrated in developmental, cellular, or pathological models for AR action remain largely unresolved. Another important area will be identifying the pathways involved in rapid signalling by testosterone, including isolating and cloning of a putative cell surface receptor.

Recent years has seen a significant increase in our understanding of the cell-specific actions of testosterone and AR. This is the result of cell-type specific knockouts of the AR gene, coupled with knockdown studies in cell culture and the use of genome-wide analyses to identify receptor binding sites and transcript profiling of androgen-regulated genes. It is reasonable to expect that as the number of Cre reporters under the control of cell-specific promoters increases it will be possible to knock out the AR gene in an ever more precise temporal and spatial manner to establish the role of AR signaling in all body systems.

Advances in chromatin immunoprecipitation (ChIP) coupled with microarray (chip) or sequencing analysis of transcripts has resulted in an explosion of data on receptor regulatory sequences and target genes. Potentially, such analysis should provide valuable information about AR regulated gene networks in different cell types and in diseases such as PCa. However, often attempts to compare and analyze data sets from different laboratories ends in frustration. Also, the reliance on synthetic androgens, while of practical value, may mask differences in gene expression for natural ligands, testosterone and DHT. It is reasonable to expect an increasing number of genome-wide studies using both cell culture models and tissues in the future, as the technology becomes ever more accessible. However, equally important will be the developments in analysis software that will help unravel the terabytes of information each experiment generates so that a clearer picture of what genes are switched on or off by the AR under different physiological conditions emerges.

The AR remains an important drug target in the treatment and management of prostate cancer, while testosterone replacement therapy is necessary in hypogondal patients. The challenge is therefore to develop and bring to the clinic SARMs with defined tissue-specific responses and minimal adverse side effects. Treatment of PCa requires better biomarkers which would allow the identification of indolent from aggressive tumors, this would allow a more 'personalized' approach to patient treatment and would also avoid unnecessary treatments for individuals whose tumor will never become life threatening. In addition, it is likely that new inhibitors for the AR protein will be identified based on our growing knowledge of receptor structure and signaling, which will ultimately provide novel drugs to complement or replace existing antiandrogens. The need for the development of new inhibitors is clear given the emergence of CRPC with current treatments.

It is also clear that there are a number of clinical situations (e.g., delayed puberty, hypogonadism) in which androgen therapy may be necessary. And notwithstanding the work of Brown-Séquard and colleagues over 100 years ago, there remain areas for which the need for testosterone therapy is less clearly defined (e.g., testosterone in the aging male). Further studies are required to generate an evidence base for the use of androgens for these indications. Currently, the majority of available therapies for androgen replacement involve the use of exogenous testosterone, however, increased use of targeted therapies that reflect the mechanism of the underlying pathology, reduce side effects, and preserve important functions (e.g. fertility) are likely to become available in the future.

In conclusion, androgens, including testosterone, are implicated in the development and function of a wide range of body systems, and throughout life, from embryonic development and masculinization, through to promotion of lifelong health and homeostasis, and into pathology and disease. Ultimately, testosterone has a single overriding role, it is essential for development of male gametes to ensure continued survival of the species. Surely there can be no greater responsibility in a sexually reproducing organism than this.